THE
WHITE
ROAD

and other stories

TANIA
HERSHMAN

SALT

CAMBRIDGE

PUBLISHED BY SALT PUBLISHING
14a High Street, Fulbourn, Cambridge CB21 5DH United Kingdom

© Tania Hershman, 2008

The right of Tania Hershman to be identified as the
author of this work has been asserted by her in accordance
with Section 77 of the Copyright, Designs and Patents Act 1988.

First published 2008

Printed in Great Britain by the MPG Books Group, Bodmin and King's Lynn

Typeset in Swift 10 / 12

ISBN 978 1 84471 475 9 paperback

Salt Publishing Ltd gratefully acknowledges
the financial assistance of Arts Council England

1 3 5 7 9 8 6 4 2

To James

Contents

The White Road	1
Heavy Bones	9
Self Raising	11
The Hand	17
Space Fright	18
I am A Camera	25
On a Roll	26
Mugs	33
Sunspots	35
Go Away	39
Brewing a storm	41
The Angel in the Car Park	53
Evie and the Arfids	56
Flora comes back	74
Rainstiffness	76
Heart	84
Knotted	85
Express	87
Firsts	92
Exchange Rates	93
Drizzling	99
Plaits	101
The Incredible Exploding Victor	103
You'll Know	115
My Name is Henry	118
Fish-Filled Sea	125
North Cold	127
Acknowledgements	135

THE WHITE ROAD

What's long, white, and very, very cold? The road to the South Pole is nearing completion . . . this road will stretch for more than 1600 kilometres across some of the most inhospitable terrain in the world.

—'The Highway at the End of the World'
New Scientist, 7 February 2004

Today is one of them really and truly cold days. You're probably thinking cold is cold is cold, either everything's frosty or you're sipping margaritas by the pool in Florida, but let me tell you, there are degrees of freezing. New York got pretty cold in the wintertime, especially for a southern gal. But all the way down here by the Pole, Antarctic minus forty ain't the same as Antarctic minus twenty-five. You need damn hot coffee in both, that's true, you got me there, but there's a different smell to the air, believe me. When I open up for business in the morning of a minus forty, I stand on the doorstep and sniff, with Fluff beside me. I say, Fluff, it's a damn cold one today and she barks, clever damn dog. Then I turn the sign from

'Closed' to 'Open' and set the water boiling for the first lot, who won't be too far down the White Road.

That's what we call it, because that's what it is, all white. Some days, you've got to wear those special glasses that they gave out on the Induction Day. Two pairs, in case one got broke. They said, Don't look at that snow when it's sunshining or we'll be putting the patch over your eyes, and that'll be enough seeing for you.

Some things the eye shouldn't see. No, some things are just too much for it.

Last Wednesday was one of them sunny days they were talking about. It was a real busy morning. I saw the first ants coming down the road around seven am. That's what I call them, Ants, 'cause that's what they are at first. I'm looking out through my big glass windows, the ones with the special coating on so they never freeze or get misty with all the heat inside. It's like you're watching a big white TV screen, it's all nothing, nothing, nothing—and then, sudden like, little dots appear: the Ants. They get bigger and bigger, and soon you see them, heading straight for me and my coffee machine. Big red trucks, with all their fancy equipment they carry to the research guys at the Pole a couple of miles past us.

It's probably Phil and Eric, I was thinking, and yeah, they pulled up and stomped in through the snow, stamping their big feet all over the floor, rubbing their hands.

'Whassup, Mags,' shouts one of them, Eric or Phil, never could quite tell the difference.

'Cold, boys?' I ask, same as I always do on a Wednesday when they make their run.

'Freeze ya soon as look at ya,' says the other one, getting stuck trying to pull his snow jacket over his big head.

'Coffee?' I say.

'You're the best, Mags,' they say together, and while they're arranging themselves in a booth, I start the pouring and bring over the cups and a couple of menus.

When I first started, half a year ago, it was quiet; everybody was just wanting to speed down that White Road and get to where it was they were going. But then slowly, they take notice of me and Fluff and our little sign for 'Last Stop Coffee', and they start coming in and making our acquaintance. They find us pretty friendly, the coffee's hot and not too bad, and I make the best damn scrambled in about a thousand white miles. I add things to my menu now and again, depending on the supplies I get through once a month when Les brings me a truckload. Sometimes it's fruit he brings me, he got hold of a box of mangos once and you should've heard how everyone was over my mango and sweet potato pie, they just loved it. Sometimes it's nothing more exciting than a whole truckload of tuna and I get to see all the different dishes I can make out of that. I can get pretty inventive with what Les hauls down here. I always was good in the kitchen, my kids'll tell you that, if you can find them. The one who's gone, he loved my scrambled the most. Ate it before it touched the plate, I used to say.

Back to last Wednesday. 'What'll it be,' I'm asking Phil and Eric. They *umm* and *ahh* and stare at the menus like they ain't never seen them before, like this ain't the only

[3]

place for hundreds of miles and they haven't been coming here and eating my food once a week for I don't know how long.

I love doing this, chatting and feeding the hungry. In between one lot and another, Fluff and I'll sit down for a breather, me with my thirty-third coffee of the day most probably, and we'll stare out into the white. You could get lost in all that white. I never knew an outside could look so clean. I thought before I got here that I would miss the colours, the greens and the blues, the yellows and the browns. Not red. I would never miss red.

But I don't miss a thing.

In the evenings we'll watch the TV. We get so many stations on that satellite, my fingers hurt from all that channel-spinning. Fluff'll bark if I do it too much, gives her a headache. She barks and I stop right there on that channel and we watch some soap opera with guys with square chins and names like 'Ridge', or a bit of the news from the real world, all them disasters and stuff. Then we hit the hay, early to most folks, but we get up when the sun does. I don't mind it, I always was an early bird. Don't want to waste your life, I told my young 'uns, but they didn't listen. Never do. Then, before you know it, it's too late.

Phil, or maybe it's Eric, asks for waffles and maple syrup, and the other one wants toast and jam, and they both drink the coffee like it's coming off the trees tomorrow and that's the end of it.

So I go back into the kitchen and set about it. I stand in front of the toaster and I close my eyes. I reach with my left hand and feel about on the counter top until I find the bread bag. I grab it and take out two slices with my right, put the bag down, trying to picture in my head where it is, and feel over to the toaster. Toast goes in first time! It's because I've been practicing. For about two months, I've been practicing with my eyes closed, a little every day. Now I can do it. I know where everything is.

It was hard at first. I dropped things, I cheated and opened my eyes to clean up eggs and stuff that slid through my fingers. I put the grill on the wrong settings, nearly burned us down, or left things so raw they could walk. But now I got it down, I can do it.

I take Phil and Eric their food, and while they dig in, I sit at the next table and we chat for a bit.

'We got two tons of gloves today,' they say. 'I don't know what they do down there, all those rubber gloves. Boxes and boxes of them. Some cutting up of stuff, I bet.'

'What else you got,' I ask, sipping my coffee.

'The week's newspapers, like always,' they say. 'Bit old now, but they get so excited when we come in. Doc Baxter, he does all the crosswords. Those guys, they're real smart.'

'They're doing important work,' I say. 'Got to have someone in these out of the way places, learning about what's going on, increasing the world's know-how, don't you?'

They nod at me, grin, stuff food in their mouths. Few minutes later, they're pulling their layers back on, paying the cheque, and out the door.

The rest of Wednesday morning people are streaming in: different delivery guys, like always, some regulars, some new, all needing serious coffee. And something special: a group of young scientists on their way for a visit to the Pole. One of the boys, he looks so much like . . . I have to stop myself going over and saying, Hey . . .

That's when I know. It's a sign. This is the day.

The afternoon was quiet. Anyone who comes down here comes through real early, in case the weather starts with its howling and rough stuff. The sun was out, it's one of them days they told me about. Dazzling, spreading light all over the white.

'It's time,' I say to Fluff. She's real quiet, smart dog. I put on my glasses, snap on her leash, open the door and we step out.

It still amazes me, like it did the first time. I don't think a body would ever get used to it, the soft clean cotton-wool of it all, stretching on and on and on. The road don't cut through it, it's part of it, just flattened out a bit. A different white, a little dirty from the cars, but not so that it gets in the way of the beautifulness of it all. I cried the first day I got here. It was like I thought peace would be.

Fluff is stood by me, her head resting next to my knee. I move a few steps towards the sun, making sure I know where the door is that I just came out of.

'It's OK,' I say to her. 'We can do this. It'll still be me. You know that.' I bend down, take hold of her leash, and straighten up. Then I take off my glasses.

At first I see everything so sharp. The white looks like gold. My eyes see little bits of gold shining all over the ground, and then it starts moving, like fishes swimming in and out of my head. Then the blurring begins. I'm dizzy, there's a pain behind my eyes, but I keep on staring. I'm not going to shut them until it's done.

I don't know how long I stand there. Slowly, slowly, someone is dropping a cloth over me and this mist comes down in front of my eyes.

Then it's all over. And it's all just white.

That was Wednesday and I have to tell you, I'm pretty used to it already. That sure happened quick. I had thought, when the idea came to me, three months after finding Josh like that and everything, that it'd be a shock to the system. Not-seeing sounded so different, like another world. But five days of whiteness and it already feels comfortable, like home. Sure I move around a little slow, with Fluff always there, giving little barks and rubbing up against me. She leads me around, pushes me in the right direction, makes sure nothing burns. She's a better person than some humans, that dog.

At first, everyone was real taken aback. I couldn't see their faces but I could hear it clear as day. But, you know, they didn't ask too many questions, and I didn't offer any answers anyway. I think most of them knew my story, about the blood, the bits blown open, the staring dead eyes, the things that I saw, things no one in this life should see. I think they heard, the way people hear everything, nothing spreading faster than a sad tale, nothing worse than a mother losing a child. Down here, everybody's got a story, everyone's got their reasons for being so far from the world. Mine's just one more to add to the pot.

Les says there's some young girl wants to come help out for a few months. Sounds good to me, she'll be mighty welcome when she gets here. But even with just me doing the serving, they keep on coming, and I keep on scrambling and dishing out the coffee.

I still sit and watch for them, only now I don't see the Ants, I hear them. It's not so different really. It's just very white, and that's the way I like it.

HEAVY BONES

'It's me bones,' I say. 'They're real heavy, I've always been like that. Honest, it's not you.' But he just stands there looking all washed out. Only a few minutes ago, we were still tipsy from the bubbly at the reception, our heads fizzing, and now I'm standing here freezing on the doorstep in my big white dress and he's looking like he's failed his first big husbandly duty and what does that say about all the rest of it and why don't we just call it quits right now. I sigh real loudly, look up and down the street a bit, rubbing me arms warm, but he's just staring into space and looking like he might cry, my skinny new hubby, with all of him drained away.

Suddenly I know what to do. I grab him under the armpits and heave him over the doorway. 'How's that, eh?' I say, puffing and sweating under all my frothy meringue. He's shocked, staring at me boggle-eyed. Then he grins and once he starts he don't stop grinning. 'Not bad,' he says, 'not bad at all, wifey.' And he plants a big one on me, right on me lips, with all the neighbours who think I don't know they're there, watching, going Oooh, look at that, bit cheeky, eh. He pulls me right in and slams the front door in all their nosey faces. 'Last one to the bedroom's got heavy bones!' he shouts. I pick up my skirts and start running.

SELF RAISING

Coronal mass ejections . . . are billion-tonne balls of plasma spat out by the Sun. On arrival at Earth they can damage satellites, disrupt electrical power grids and even kill astronauts.

—'3D Space-Weather Forecasts on the Horizon'
New Scientist, 4 November 2006

I make them out of flour, sugar, eggs, like you would any cake. But they're not any cake, they're lab coats and test tubes, DNA and petri dishes, just like in Science at school, when I used to get things right and the teacher would say, Excellent, Madeleine, that's exactly what happens when magnesium oxidises, and he'd smile at me and I'd grin at him, and the rest of them'd laugh and throw things and call me Swot. But I didn't mind about the names they called me; I knew what I knew and I wasn't going to pretend I didn't.

The Lab Coat: chocolate cake, marzipan, chocolate buttons. Marvellous, they said to me, sent photos from the party, showed the professor cutting it, cutting my cake, all the scientists standing, watching, while he's slicing into it, right down the middle, right through its heart.

I kept on going, onto University, more stirring and mixing in the labs and learning about neutrinos streaming from the Sun, about quarks and electrons and protons, charge and spin, nuclear fusion, potential energy, about stars and supernovas. I was loving it, but then I met John. He was studying engineering, it all happened quickly, he proposed, I said Yes. What did I know? Didn't know then that I'd get pregnant right away, that I'd have to drop out, always meaning to get back, always meaning to. But after the first one, we had two more, and it turned into me snatching a bit of time here and there, turning on Radio 4 and maybe getting five minutes of the science program on a Sunday, or a quick look at the newspaper, see what's up with new particles being found or global warming or gene splicing. Going back to university became like a dream, something I sort of remembered but couldn't really see anymore.

DNA: column of vanilla sponge standing on a base, dyed strands of orange peel, one green and one red, wound around it. Graduation party, PhD in molecular biology, they told me, and I pictured her in that cap and gown, beaming, knowing so much about what goes on in our cells, her with time to learn and learn and keep learning.

I loved the kids, 'course I did. They were little monsters some of the time, but I enjoyed myself, I really did. I tried to get them interested in what I loved, bought them chemistry sets, the girls and the boy, tried to show them about the world, how we should ask questions, look inside things, find out how it all works. It was the other mothers I couldn't stand. Never wanted to talk about anything that mattered, never interested in anything that needed a brain, just chattering about teething and schools and Oh look at her, she's crawling backwards, isn't she clever? Whoopee, don't they all do that? Yours is nothing special, I wanted to shout at them, nothing special now, but maybe she could be, maybe she could be great, if you gave her clever books and toys that got her thinking instead of dolls and hair ribbons. But I never said it. They wouldn't have known where to put themselves, and then they would've left me out.

The Test Tube: rectangular sheet of caramel wrapped into a column, covered in marzipan, little meringue balls stuck on the edge, like liquid bubbling over. At the school Science Fair, they put it on the stage, and I sat and watched little boys picking off pieces of the marzipan while everyone else was looking at the experiments. Weights hanging from pendulums, chemicals heating over Bunsen burners and turning into something else, prizes given out, parents clapping, and me sitting there, watching.

John wasn't sick for long, it spread quickly, and I sat there in the hospital and saw him leave me. I held his hand and wondered what other life I knew, after all these years. The

[13]

kids left home to go to University, medicine, architecture, engineering, getting lives of their own, and me now with time to listen to Radio 4 for hours, time to read every Science page in every newspaper. But all I wanted was have him back, grumbling at me from behind the Sports section, drying up what I'd washed, tickling my feet when we watched telly. I'd have given anything.

The Petri Dish: round cake iced in white with a raised edge, and tiny coloured marzipan shapes, red and green and blue, cells swimming around. Some professor's retirement party, he almost won a Nobel Prize it says on the Internet. Dedicated his life to science. His wife and kids probably never saw him, Daddy's at the lab again, but proud, so proud.

I made cakes because I didn't know how to do anything else. I might remember how magnesium oxidises, but that's it. I can bring up kids so they turn out all right, but I'm no scientist, not any more. It's all too late. So I set up Lab Cakes, got myself little business cards and stuck them on the University notice boards, put ads in the local paper. I keep myself busy. Everyone oohs and aahs, but it's not hard, not really, when you've got all the time in the world to think about it. Anyone could do it.

The Sun: chocolate cake ball made in Christmas pudding mould, orange icing with brown smudges for sunspots, angel hair spaghetti mesh for the solar clouds, blue-dyed pasta as plasma shooting out from the solar storm.

The minute I step out of the lift and those smells hit me, I feel at home. I walk down the corridor past all the labs, wheeling my shopping trolley, and I take deep sniffs and part of me wants to push open a door, peek inside, shake a test tube, look into a microscope. But I don't. They don't want me in there, messing with stuff. I just keep pushing the trolley with my cake inside it until I get to the Common Room.

The guy is there, Dr Williams, who ordered it for the party. He shakes my hand, shows me the table where he wants it, and helps me lift the box. Then I get the cake out very gently and set it down on the table and I turn around and when I see his face I can't stop myself from grinning. My goodness! he says, and his mouth's dropped open, he's mumbling something about sunspots and solar clouds. I know what he's talking about. He sent me pictures and links to the Internet and I read more, about coronal mass ejections and plasma. I start to say something, I've got some questions, I want to ask him about his research. But he's just staring at the cake, he's forgotten about me, so I push the trolley back out of the door and down the corridor to the lift.

When I get there, I'm about to press the button when I stop. I can see him in my head. There he is, with all the other scientists, and they're watching while he takes his knife and cuts into my Sun, plunging that blade all the way through, solar clouds splitting into tiny bits, orange icing falling all over the table, chocolate cake bleeding crumbs. I let go of the trolley, my heart is beating like its jumping out of my skin, and I turn round and walk fast, and then I'm running back through the Common Room door.

He's still sitting there, gazing at the cake like he's in a daze, and when I grab it he stands up, says, Hey . . . But I'm over at the window, it's already open, and before he can stop me I lean out and with both hands I throw my Sun as far into the sky as I can. And when, instead of doing what gravity says it should, it floats up and up, orange and brown, dancing and spinning, into the clouds, I'm not surprised one bit. I watch it climb higher and higher until it's just a speck. And then it's gone.

THE HAND

Her elbow twitches. He doesn't know her, her father, her community. He doesn't know that her long skirt, long sleeves, means that she doesn't, can't . . .

His hand floats between them.

Will you be warm, soft, cool, moist, strong? Will you take mine gently like Rivky on the way to school? Or will you be firm, squeezing, crushing? When our skins touch, will I jump, gasp out loud? Will you know that I haven't . . . ever?

And afterwards: will you be printed into my palm, an impression in clay?

Elbow twitches, wrist jerks, and her fingers move stiffly into the air, reaching for his.

SPACE FRIGHT

Daring aerospace entrepreneurs race for the X Prize, a $10 million reward for the first private vehicle to fly passengers to space . . .

—'Space Fright'
New Scientist, February 14th 2004

'I've heard of men being hard to pin down,' said Agnes, 'but this is ridiculous. Didn't you read the gravity section in the manual?'

Bill floated helplessly above her.

'I'm really sorry . . .' he said from the ceiling. 'I had no idea that it would be so difficult to, well, you know, keep one's feet on the floor, so to speak. The sidespin is intended to compensate for the, umm, gravitational deficiencies, but of course I should have activated the trip-switch to alter the undercraft sensor system. I'm sorry, I just don't know why I didn't. It all happened so fast.'

He looked so miserable that Agnes smiled broadly to cheer him up. She gripped the two handles on either side of her on the wall that she had flung herself against when

Bill had decided they could undo their seatbelts without having first checked the settings. In this position, with her arms pinned back against a control panel, she felt rather exposed. But then, on the positive side, Bill was getting a good view of her breasts.

Well, thought Agnes, this was certainly different from other dating adventures she had had: Nanospeed Night, where you had two seconds to make your mind up about a man before he moved on and was replaced by number twelve out of two hundred suitors; the Dinner in Space Darkness fiasco, where you were supposed to interact with various men unprejudiced by their facial drawbacks, around a table that none of you could see, trying to eat food when you had no idea where the plate was. She had apologized profusely to the man on her right, but wasn't sure he had heard her as he was rushed away screaming, taking her fork with him. She hadn't met anyone that night either.

With his head grazing the ceiling, Bill felt like he might cry. How long had it taken him to get this woman—any woman, for pity's sake—to take a spin in his new XCOR 5000, which had extra comfort features and a dual spin turbo backdrift with built-in stabilisers; how many times had he run through his space seduction scenario ('look at that view of the cosmos'—slide arm around shoulders; 'doesn't it make you feel small and insignificant?'—go in for the kiss)? And now here he was, suspended in an awfully unmanly way several feet above her, unable to get to the bottle of perfectly chilled Australian white in the mini-fridge (a new feature that many had complained earlier XCORs were sorely lacking), too far from the stereo button to even mellow the situation with a little Van the Man. He would have whacked his head against the wall in despair, if he had any control over it his body at all. Well, he was getting a good look at Agnes' untouchable bosom, at least.

Agnes was staring out of a porthole at Earth.

'I think I can see my house,' she said.

'Blimey,' said Bill, 'you have great eyesight.'

Agnes liked this about Bill. He was quick with a compliment, but not in the sleazy way of some of her dates, whose praise seemed to ooze. Bill was rougher, it was true, and he certainly didn't know when to stop talking about mind-numbing technologies and how they functioned on a molecular level, but he had a certain inelegant persuasiveness, otherwise how would he have managed to get her here? Agnes always believed she had vertigo, but staring out of the porthole, she felt a great calm.

Bill, on the other hand, was becoming increasingly frustrated.

'Agnes,' he said, in what was almost a wail. 'Could you help me down somehow?'

Agnes was transfixed by Earth.

'Agnes?' Bill said, a little louder. 'Agnes!'

'Oh, sorry,' said Agnes. 'What?'

After some discussion and much cajoling on Bill's part, Agnes, keeping tight hold of the handle on her right, let go of the left and tried to propel her body towards the control desk. On first go, she started moving upwards.

'No, stop!' yelled Bill, and Agnes managed to grab hold of a red lever.

'Umm, I don't think you should pull that.'

'OK, but what did you want me to do, when you yelled "Stop!"? I was going up and I would have been . . .'

'Yes, but you just can't grab any old . . .'

Agnes let go of the lever, turned away and stared out through the porthole again. Men, who did they think they were? He was stuck on the ceiling, he was the one who screwed up, she was just trying to help, and then he starts shouting. She might have sniffled a little, but she couldn't get to her tissues to wipe her nose, so she didn't bother.

Some time passed. Agnes lost herself in contemplation of the Earth's surface, its lands and seas, drifting into a kind of meditative state where she was floating outside the XCOR, free from the shackles of existence, liberated from the tyranny of traffic lights, credit card bills, her mother, bad dates.

Bill felt like he might have a temperature. Could being upside down endanger your health? He had read last week that scientists were worried about new types of blood clots from too much space travel, but their trials weren't conclusive so he tried not to panic. He had cramp in his right leg, which was actually above the rest of his body. This was pretty much the worst date he had ever been on. It beat the one where after ten minutes in the restaurant, before they had even ordered, she got up, told him it wasn't going to work, and left. Bill had stayed, shocked by the speed of the date's death, drowned his sorrows in three shandies, and staggered home, where he spent the evening watching *Friends* re-runs.

He looked across at Agnes. What the hell was she staring at? It's not like the view changed much, never-ending blackness apart from the VZ2 constellation and several minor stars, and Earth, which wasn't moving at all. He was stuck on the bloody ceiling. How unsympathetic could a woman be? Weren't they meant to be the ones that felt emotions and helped people? Why had he brought her? So what if Agnes was the best-looking female in the pottery class that he had been going to every week not to make stupid vases but because it was supposed to be the ideal pulling venue? She was weird, but he had thought that meant she was nice-weird, not odd-spooky-weird. She was certainly no bloody use to him now.

Bill looked around for the best handhold, located a small shelf and a safety belt strap he could use for manoeuvre, and began a slow scramble, head first, down the wall opposite Agnes. Agnes, lost in her dream of a peopleless Nirvana state, didn't notice anything. By dint of some inventive moves, Bill managed to get quite far down the wall, head first. Then he decided he had better right himself in case too much blood to the head did something nasty.

'OK, here we go,' he muttered. 'Come on, come on.'

With one almighty push he shoved his legs from above him and swung them over his body.

Agnes awoke from her reveries with a start as a foot struck her left breast. Instinctively, she grabbed it.

'What are you doing?'

'What do you think?'

'What should I do with . . . this?' She nodded towards his foot.

Bill paused. Then he very slowly said, 'Perhaps you could lower it down to the floor, so I could stand up? Could you? Do you think you . . .'

Agnes bent as far as she could with one hand still holding the handle. She placed Bill's foot onto the floor.

'There,' she said, wishing that he wasn't.

'Thank you.'

A bit more struggling and much use of his stomach muscles brought Bill upright. Now that he was vertical, he felt rather more kindly towards his date.

'Much better,' he said loudly. 'Well, how about I find that white wine. I might just be able to get over to the fridge. Okey dokey, let's see, there's something over there I could hold on to, you just sit tight. Well, you aren't really going anywhere are you!' He laughed a little and then stopped. Agnes said nothing. She was staring at him.

Bill had no idea what her expression meant.

He had no idea that inside Agnes a malevolent glee was taking hold, working its way up her body towards her brain.

Agnes was fed up with people, with men, and especially with incompetent, disorganised, unromantic, breast-staring, space-mad, technology-obsessed Bill.

Bill watched in horror as she let go of the handle with her right hand and grabbed with two hands the turbo-thrust lever he'd told her on no account to touch. Agnes pulled it down with all her might. It was as if Bill was watching a slow motion sequence in a film, then it felt more like the spin cycle. The XCOR began to rotate, slowly at first, then faster.

'What've ya dooooooone . . .' Bill tried to speak as he spun.

'Ooooooops,' said Agnes happily. Some dates were so boring, at least this one had a bit of momentum.

I AM A CAMERA

'I am a camera,' she whispered to herself in the shower, sliding her fingers along the rail already installed for the day when she wouldn't be able to find her way out. She thought of herself as one of those old devices with a photographer hiding under a cloth, producing sepia-washed pictures. She clicked and whirred and stored images inside, cataloguing scenes from her memory. Faces and landscapes, each titled for easy access, later on.

When the day came, the shutters of her camera floundered against the darkness. She sat still, noises pressing around her, and opened her photo album. 'May 1st, Brighton, me and Simon,' she murmured, and her inside eye saw the colours and textures. Images came up as she bid them. 'Stay with me,' she whispered, the pixels dancing on the inside of her eyelids. She gripped the arms of the chair. 'Stay with me,' she demanded, but the colours were already starting to fade.

ON A ROLL

*Understand randomness and you could win a Nobel prize, or clean
up big time at the casino . . .*

—'On a Roll'
New Scientist, 6th November 1999

Holding them by the heels, I set the sandals down on the
cloth. My naked toes twitch and wriggle. The croupier's
expression doesn't flicker, as if women bet their shoes
during every roulette game.

'Six hundred and forty-five dollars,' I whisper. The
croupier, the waitress with the tongue stud, Jim from
Texas—they can all see the label. They know what these
golden babies are worth. 'Brand new. I have the receipt,' I
say, braver now. This is the way it is meant to be. After all
I've been through, right here is where everything turns
around.

'Mind the heels on the baize, ma'am,' says the croupier.
He pushes over my chips, and, even though I know what

he is about to do, I can't breathe as he picks up my beautiful sandals and slides them out of my reach.

I count my chips: three blood-red hundreds, six night-blue fifties, four bile-yellow tens and a pitch-black five. My toes grip the legs of the high chair, my hands are shaking. But there is also something reassuring about all of this. It's word perfect. Everyone is playing their part so I am playing mine.

Las Vegas was a business trip. The company sent me to meet a client, some old bloke with a plastics factory churning out casino chips. I was there to walk him through the new laser slicer. I arranged a stopover in New York, six hours. That was all I needed.

It was a night flight. I was exhausted, hadn't had an uninterrupted night since . . . I nodded off even before we left the ground. I woke up suddenly when something bumped my seat, confused, still inside the surreal dream I'd been having. A roulette table. Jim from Austin. A croupier with enormous eyebrows. A waitress, heavily made-up, tongue stud. And me, in my favourite green dress. And the shoes. The shoes my stop in New York was for. I sit there, in my green dress and my gleaming sandals, and I gamble away all my money.

In this dream, people didn't appear out of thin air or have two heads. I didn't sprout wings and fly. It all felt so real. I sat there on the plane replaying it in my head, something in me wanting to cry like a baby.

I came out of JFK and found a taxi. 'Bergdorf Goodman's, please,' I said to the driver, and stared out of the window at the city I'd only seen in films.

I headed straight in and up to the shoe department, trailing my wheelie case behind me.

There they were.

I asked for my size and whispered to the butterflies in my stomach. Slipping the straps over my ankles, the little Roman coins gently knocking into one another, I had to remind myself to breathe. My feet, always ignored, shoved into ugly leather clodhoppers, were suddenly elegant, pale, long. Someone else's feet.

I could have picked a glitzier style with rhinestones, or something plainer and more sophisticated, but it was the coins. The coins reminded me of a museum we'd visited when I was a kid. I'd wandered past the perfume jars and animal skins, but stopped short at the case with ancient money. People thousands of years ago had used these bent bits of metal to go shopping? What I think I found the most strange was that all these lives had gone on before I was born. Until that moment, I thought the world began and ended with me.

I paid for the shoes and went to catch my next flight.

In the hotel room I only had time to change my clothes and have a quick cup of tea before heading off to my meeting. It went well, he seemed to like me, touching my arm every few minutes as only an aged flirt can get away with, promising he'd upgrade all their machines by the end of the year. I refused his dinner invitation. Roulette wheels were spinning in my mind.

Opening the hotel room door, the silence hit me like lead. I stood, frozen, in the doorway. No one was waiting for me. No one would ever be waiting for me. A fit of hysteria rose inside me. But something stronger forced it down. Not now. I couldn't lose it now.

Taking the green silk dress off its hanger, I slid into it. I opened the Bergdorf shoe box, closed my eyes for a minute, and then lifted out the right sandal, tiny coins jangling.

I went to the nearest casino, a tacky imitation Greek temple, white pillars and nude statues. High on my heels and dizzy with it all, I bought two hundred dollars-worth of chips and looked around for the roulette tables.

And there they were.

A croupier with bushy eyebrows was collecting chips from a balding man sitting with a glass of whisky. Something whispered in my chest. Someone was pulling my strings and I let them lead me.

I sat down to the croupier's left, opposite the man who would soon be telling me the story of his three marriages.

'Can I get you a drink, miss?'

Thick kohl eyeliner, red lipstick, short dark hair. Tongue stud. I ordered a Cosmopolitan.

'Are you in, ma'am?' said the croupier.

'Yes,' I said. 'I'm in.'

'A Brit,' said the balding man, leaning towards me with a leer.

'How can you tell?' I said, smiling. I'd been here before.

It took me an hour to double my money and then lose it all. I picked red and black at whim, high and low. I trusted I was doing what I was supposed to.

Jim was down a thousand dollars.

'I just can't hold back,' he said. 'I just like to watch the pretty little numbers going round.'

I stared at the place where my chips had been.

'Out?' asked the croupier, turning away as if he already knew the answer.

I could have stopped. I could have walked away with nothing but an hour's fun for two hundred dollars. I could have sliced through the strings that had my hands dancing and my tongue moving, and gone back to the hotel and straight to bed.

I reached down to undo the strap on my right shoe.

The croupier stands, statue-like, my beautiful sandals by his right hand, waiting.

'Well, girl, you gonna do it, blow the lot?' grins Jim. He drains his fifth scotch, watching me.

The waitress whistles softly.

'Honey, I'm not supposed to interfere, but those little darlings, you don't want to just . . . I mean, you can't . . .'

I barely hear them. I take hold of the pile of plastic discs with both hands and push them away from me onto red 32. Thirty two. The years of my life. And for what? A husband buried before he had a chance to get old. I thought we had all the time in the world, but all we had was an instant.

All he had ever talked about was having children, but he never knew about the one he almost had. We'd only just got engaged, we were barely beginning. Now I have nothing left but a wardrobe of T-shirts and jeans and his favourite boots.

I look away from the pile of plastic. Folding my hands in my lap, I say: 'Roll,' and the ball begins its journey. Fast at first, it circles the numbers, bumping and jumping from one to another. My eyes can't keep up with its dance. Then it begins to tire, moving around the wheel more lazily, looking for its place. Slower and slower it winds, over ones and twenties, fives and fifteens, and then into thirty-two it sinks as if it has come home.

And then out.

I lose.

I lose six hundred and forty-five dollars. And I lose my shoes.

The croupier looks at me, blinks.

'I'm sorry, ma'am,' he says. He gathers the chips to him, and then he slides my sandals into a drawer under the table.

The waitress puts her arm around my shoulders, the Texan offers me a handkerchief. They expect me to cry, to beg and plead with the croupier, to claim that this is unfair, that betting a pair of shoes must surely be against the rules, to threaten to take it to the management.

Instead, I stand.

'Thank you,' I say to the croupier. He nods. Jim winks, wishes me luck.

The waitress picks up her tray and takes away my cocktail glass.

My toes sink into the red carpet.

'A good night, ma'am?' says the security guard on the door. Then he sees my feet. His eyes open wide. 'Miss, are you . . . ?'

I am out in the street, barefoot. People are staring but I don't care. I am free. I have sacrificed the most perfect things I ever owned, have shed my skin and am standing naked and new again. I am Eve in the garden before the snake arrives, and this time I am going to do it right.

MUGS

They meet in pottery class. Her coffee mugs are misshapen, clumsy. His espresso cups are identical, boring. He envies her creativity; she craves his perfection.

After class, they walk to the bus stop. Shy, hands in pockets.

'Do you . . . ?'

'Yes?'

'I mean . . . are you . . . ?'

'Am I . . . ?'

'Hungry?' They laugh, relieved.

The air is warm with tomato sauce. They order gnocchi, spinach ravioli.

'My mother wanted a boy, made me wear trousers when everyone else was in frilly dresses,' she says, after two glasses of house red.

'My brother shut me in the wardrobe for hours,' he says, looking down at the plastic flowered tablecloth. She shifts her hand so her fingertips touch his, just for a second.

At the end of the final class, they pack their work, carry heavy bags towards the bus stop.

'Hang on,' he says. He steers her past and along, to a dark alley beside the bank. Nervous, she follows. Putting down the bag, he gets out a parcel, unwraps the newspaper. Then, like lightning, throws it at the wall.

Bang! A thousand tiny white splinters.

She stares, amazed. Then laughs. Laughs until she cries. He grins, watches as she scrambles to unwrap one of hers and then, a bowler, swings her arm and flings.

A shower of clay handles and coloured chunks.

He throws another, and then she, and they take turns until there is only one left of each.

'Please,' she says. She takes his smooth espresso cup, cradling it. He holds her clumsy mug to his heart. They stand in a pool of pottery pieces in a dark alley, looking at one another, in a city of a thousand sighs and lonely souls.

SUNSPOTS

A study . . . has found a link between the cost of wheat in medieval England and the mysterious sunspot cycle.

— 'Solar Cycles Drove Medieval Markets'
New Scientist, 20 December 2003

He knows not of what he speaks, says my husband. *He has not been right for many a year, you know he talks nothings.* He takes my arm, to lead me away from the marketplace but I do not move.

The child will not live, the simpleton says again. He looks at me. Points to my belly. *He will shrivel. He will wither. Look to the sun. Her dark places. You must count them and beware.* I bow my head. The sun greedily burns my pale neck as I allow John to lead me away. The simpleton calls out to me. *The sun has her dark places. You must count them, count them. Let him go.* Then he laughs a terrible laugh and my heart eats itself in fear.

He is a fine healthy baby, says the doctor who comes when John calls for him. *You must not fret yourself, you are a strong woman, be getting on with your duties, there is nothing to fear.* John smiles, holds my hand, thanks the doctor. I felt nothing but a chill in my spine and a tumbling in my stomach.

At night I get up. My husband is asleep and happy in his slumber. I go out into the field wearing only my night-gown. I wet my feet on the damp grass. My hair hangs plaited down to my waist. My belly shifts the material forwards, higher at front than back. *Shhh,* I whisper to my child. The moon looks down. Kind moon, protecting me from her sister sun.

At the top of the hill I lower myself to the ground, my back against the tree. *This will be yours,* I whisper to him. *You will live long and strong. Your bones will not snap, your neck will hold firm. He does not know anything. He is a simpleton. He speaks only nonsense. You will be my perfect boy. Golden skin and full of laughter like your father, dark hair and wise like your mother.* The child shifts in my belly. I know he hears me. *Stay safe inside me. You will know when it is right.* The moon shines silver and I pray this night will last for ever and the sun will not rise.

I awake in the morning unrested. John does not know I have been out. I am careful to hide from him the damp upon my nightdress, the night air in my breath. He is cheerful, as is his way. I cry inside as he whistles and sings. I heat the water and make bread, and into the dough I pour my fears and the stirrings of my heart. I bake the bread but I burn it and it is wasted. I throw it outside for the pigs. It was not good bread, it had my fear for binding. The pigs will not touch it.

We pass more nights and days, the sun growing in her power, her dark places shifting and shaping. I see them from the side of my eye. I glance sideways and see her staring down at me, waiting to devour my child. Now I am nine moons gone and I am waiting too. They say a woman is always waiting. John does not wait. He whistles and works, kisses my mouth and sleeps in peace.

My pains come in the marketplace. I have fish in my basket and potatoes. I drop the basket and see the fish slide onto the stones. I hear a scream and it is my scream. My belly is on fire, I am splitting open. I fall to the floor, and I see the fish on the stones, a blood eye staring into my own, as I listen to my own screams and there is a heat between my legs. They come and carry me to my bed. I think I hear the simpleton. He is laughing.

You must push, they say to me. I push and I push and the pains are a blanket over my body, smothering me and I think I must die. But I do not die. The child comes from me, trailing blood. *It is a boy*, they cry. *You have a son!* I am weak. *Does he have everything*, I whisper. *Is he whole? He is beautiful*, they tell me, *a beautiful boy*. They try to hand him to me. From the corner of my eye I see him, pink and red, his head covered in my thick black hair, his arms and legs soft and tough.

I turn away. I cannot take him. *Here is your son*, they say to me. I say nothing. They mutter and shuffle and hold him and say that they do not understand me, that a mother must take her child, that I am possessed.

I feel him gone. I feel the space inside me where he laid his head. There is a hole near to my heart and the hole is the shape of him. I hear his cries and I do not know him. I close my eyes and in the darkness I see the sun burning.

GO AWAY

Knocking. It's him. I'm behind the sofa quicker than blinking. Harder now. Go away, piss off. Wearing his knuckles out and bothering my doorbell too.

I try to avoid his phonecalls, but can't keep that up for ever, so once in a while I give in.

'How are you?'

'Fine, all's well.' I try to sound cheery so he'll stop bothering me, praying in my belly he won't start in about the moron wife, the garden, his work.

He's an actuary. Like axe-murderer but without the thrills.

Few minutes later and . . .

'They've put me on this new project, really fascinating . . .' Swallowing yawns, I send my mind strolling, a beach or some foreign city, alone, nobody knowing me, while he mewls on. How did it come to this?

Knocking stops. Don't breathe.

One minute.

Two.

I straighten up, hankering for tea.

Then I hear him.

'Mum?' he's calling. 'Are you there? Mum, it's me.'

I freeze. Give in now and you're done for. An hour of torture, his stupid fat face, whining on, and you itching to grab a frying pan and end it all.

Then he's gone. Birds sing again and I'm towards the kettle.

Sometimes I think about just upping and leaving. Not very nice, eh? There's probably groups for mothers like me, crying and working out where we went wrong. That's a laugh.

'I'm not the problem,' I tell the robin on the windowsill, and I sip my tea, 'For Sale' signs dancing in my head.

BREWING A STORM

Those plagued by water shortages or extreme weather are taking matters into their own hands: schemes to enhance rain or suppress hail . . . are underway in more than 24 countries.

— 'Brewing a Storm'
New Scientist, 8th November 2003

After a breakfast Bloomfield rated as five on his ten-point scale (eggs too runny, waiter preoccupied with impressing waitress and not immediately attentive to customer's needs), Bloomfield stood in the rather shabby and unattractive brown-themed lobby of his third hotel that week, briefcase in hand, waiting for his guide. This guy was probably the translator too, just like on Bloomfield's other site visits. These Scandinavian countries, why were they still speaking their bloody languages? Might as well be Klingon. Why couldn't they all just give in and accept that English was the Mother Tongue? It would save so much time, he thought, swinging his briefcase as he imagined himself presiding over a forum at which all countries

[41]

accepted the Queen's English and saluted Bloomfield for his brilliance. We never would have dreamed, the Belgium envoy was saying to Bloomfield, presenting him with the Medal for Linguistic Liberation, that life could be so transformed. Thank you, thank you . . .

'Mr Blumfeld?'

Bloomfield blinked. A tall Aryan was standing in front of him. Blue eyes, blond hair, IKEA model, the perfect Sven. Why did they all have to be so bloody big? Bloomfield was unswerving in his conviction that five feet four inches was the ideal height, as long as all shelves and kitchen cabinets were suitably positioned. Sucking in his stomach, Bloomfield said,

'Bloomfield. Yes, that is I.'

'Gustav Gustavsson,' said IKEA man, gripping Bloomfield's hand. His English, Bloomfield noted, was completely accentless. Well, at least this chap's learned something. But Gustav Gustavsson? Bloomfield suppressed a snort. When will they get it? Gustav took Bloomfield's arm (another thing for the list: too much touching in these bloody places), gently steering him towards the automatic glass doors.

'Have a good day, Mister Blumfeld,' called the desk clerk.

Bloomfield ignored him but noted the courtesy. Not bad, he thought, not bad. Aha, the Volvo. Of course.

The car, silver and extremely shiny, was not driven by Gustav himself but by a chauffeur. That's the size of it, thought Bloomfield. Nothing less than my just deservences. After a drive of fifteen minutes out into the countryside, during which Gustav told Bloomfield a little about his childhood in a small village (rather forward though amusing) they pulled in through the gates of the facility. METEOROLOGICAL ADAPTATION RESEARCH CENTRE, Bloomfield read off the large sign above the gate. That's a new one. *Meteorological Adaptation*. Very PC, very politically bloody correct.

The main building looked brand new, all aluminium and glass. Bloomfield pulled out his Treo (pen and paper, strictly amateurish, must show locals your command of technology) and started making notes, writing down 'Meteorological Adaptation Research Centre' and 'shiny building'.

'Ah,' said Gustav. 'You have a Treo. Very good. Of course, we are loyal to Ericsson over here.' Laughing in an annoying way, Gustav whipped out the latest model that Bloomfield had been salivating over only the week before in the *Observer*'s Gadgets section: phone-plus-camera-plus-music-player-plus-PDA-plus-bloody-coffee-maker-and-cuddly-toy, thought Bloomfield, feeling his blood rising.

'Well, yes,' Bloomfield replied, breathing hard, 'but sometimes you just want the device to do one thing and one thing only. Who needs to play MP3s and watch your mother-in-law at Christmas dinner when you're trying to work, eh?' (Scored a point there, especially by using the term 'MP3'. Very hip.)

'This way,' said Gustav, pocketing the Ericsson. Bloomfield watched it disappear into Gustav's jacket. 'We are meeting Mr Henrik Johansson. He is the manager of the centre. He is very much looking forward to your examination of the centre's project.'

The lift ride, of course, was smoother than a baby's bottom. Ten flights without a whisper. Bloomfield patted his hair down and took a deep breath. Get into it man, he told himself. No emotion, all control.

Henrik Johansson was made of the same stuff as Gustav, blond and more blond. He shook Bloomfield's hand even more firmly, and gestured towards the long pine conference table. IKEA, thought Bloomfield. Don't you people have any other shops?

'Coffee?'

Bloomfield nodded, saying nothing yet. Strong and silent was his preferred initial tack. At a coffee machine in the corner, Gustav started frothing the milk.

'Welcome, Mr Bloomfield,' said Johansson, who surprised Bloomfield by sitting not opposite but beside him. He didn't seem to need a translator either. 'We are extremely happy you have come. We have something very exciting to share with you.'

'Thank you,' said Bloomfield, assuming a serious yet attentive position. Johansson clicked a button on his ultra-slim laptop, and a Power Point presentation emerged on a flat screen on the wall opposite that had been showing what Bloomfield now understood was a screen-saver and not a painting of the Swedish countryside. The presentation was titled 'Meteorological Adaptation: The Challenge of Modern Civilization'.

'Weather is the bane of our life, is it not Mr Bloomfield,' said Johansson, turning to look him straight in the eye.

'You are from that country where you can never plan an outside party, a birthday or a wedding, perhaps, even in July, in case the heavens open and you and your guests become drenched. Why do we allow ourselves to be at the mercy of the elements? Well, until now we had no choice. Oh yes—we could pray to the Sun God, am I right?'

Johansson laughed heartily, as did Gustav by Bloomfield's right ear, giving Bloomfield rather a shock as he leaned down with the cappuccino. Bloomfield smiled and then reassumed his professional, attentive expression. Not too much jokery, gentlemen, he thought. Don't have all day. Well, yes I do, but . . .

'No more gods, no more praying for the sunshine, Mr Bloomfield.' Johansson's voice was moving steadily upwards. He clicked the mouse, and the screen morphed into *Cloud Rehabilitation: The Next Generation in Meteorological Adaptation*. 'We here have been working for several years on an entirely new strategy. Cloud seeding was obviously a failure. Amazingly, some countries are still chasing the shadow of that scientific joke. No, we have a different way to do it: Cloud Rehabilitation.'

'Cloud Rehabilitation,' said Bloomfield, intrigued. A film began playing on the screen. Grey, menacing rain-clouds hung over a city which could have been where he spent last night or anywhere in Europe. The camera zoomed in on people in the street staring at the sky in horror and then hurrying off to their warm, safe homes, which glowed in a radioactive kind of way. Soon, the city was deserted. Lightning flashed and thunder rumbled. I get the point, thought Bloomfield. So?

'So, what do we do about this?' said Johansson. 'We have a device that we call a 'cloud rehabilitation encouragement machine' or CREM. We simply point it at the offending clouds, and it makes them into harmless bundles of fluff.'

He and Gustav laughed again. Bloomfield leaned forward. He couldn't wait to hear how this baby worked.

A new film started rolling on the laptop screen. A sleek, machine-gun-like device on a rooftop. Johansson moved the cursor over a button on the device's handle, clicked, and a stream of green particles shot out towards a dark and nasty rain cloud, which turned pink and then a happy cream colour.

'Voila!' said Johansson. 'A cloud is rehabilitated! Now, how do we do it?'

'Yes, indeed,' said Bloomfield. 'How do you do it, eh?' He laughed, but he was the only one.

'It is all about electric charge,' Johansson said solemnly. 'A rain cloud is not like a regular cloud. It is negatively-charged. It is the H_2O that makes it different, all that water stored up. What we do is send a stream of specially-adapted electrons — and that is our little secret — up into the cloud, and the charge is encouraged to become positive.' His face switched into a large grin. 'So, bye-bye rain cloud, hello sunshine!' he said, flinging one arm out to the side.

Bloomfield was afraid Johansson might start singing.

'Where does the water . . . err, H2O, go,' he asked, feeling that this was not a stupid question.

'A very good question,' said Johansson, clapping him on the shoulder, making Bloomfield blush. 'The H2O is turned into H3O. Do you know what this is?' Bloomfield shook his head.

'It is tri-hydro-oxidase. Completely harmless, even good for the atmosphere,' said Johansson. 'It filters the sun's rays so we do not get too sunburned on our holidays, eh?' And he laughed again, sitting back. Well, aren't you pleased with yourself, thought Bloomfield. But even he could see that the idea was clever, bloody clever. Well, let them try this one for size.

'How many tests have you done? What's the accuracy of this thing?'

'Another very excellent question!' said Johansson. 'They have sent us just the right man for this job.' Bloomfield felt himself turning red again. He made a mental note to see a doctor about his blushing problem when he got home. Johansson continued: 'We have done testing over two years, in different parts of the country, and we have had a success rate of 97 percent.'

Ninety-seven percent? Bloomfield couldn't believe his ears. A miracle. It couldn't be! This was very different to what he had heard on his other visits, where there were extremely complicated scientific processes but very few actual results. Wait until he told HQ.

After the presentation, Gustav took Bloomfield to the top floor to see the machine, although he had to stand behind a glass wall.

'I am sorry you cannot approach,' said Gustav. Bloomfield took a good, long, professional look at CREM. The device was designed like an oversized Uzi, with some kind of sun logo above a set of silver buttons. The nose tilted slightly upwards, and tapered off to a ten-pence-piece-sized hole. That's where the magic comes from, thought Bloomfield. Clever electrons. Very impressive. What a report he was going to write.

'I am sorry we cannot show you CREM in action, but today is such a lovely day,' said Gustav, gesturing to the window at the end of the corridor, awash with blue skies. He laughed and Bloomfield laughed along. They went for lunch. Bloomfield enjoyed his meatballs and started feeling extremely happy, even telling Gustav the story about his dog getting run over and how the vet said he would have to put the little guy down, but how his tail started to wag and now Boxer was as right as rain.

'As rain!' laughed Gustav. 'You English will have to change your expressions, won't you! The next generation, they won't know what rain is in their cities!' Bloomfield chortled. He could see the future, and the future was CREM and sunshine.

It was when Bloomfield went to the bathroom that the day began to change for the worse. He was sitting in a stall when something was pushed under the door.

'Hello? Excuse me . . . ,' said Bloomfield, who didn't like to be disturbed while doing what he had to do.

'Read this,' said a voice urgently. 'It is not what you think.'

'Who are you? What is this?'

Silence.

Bloomfield hurriedly finished his duties, flushed and picked up the grey folder. When he saw what it contained, he felt faint.

Bloomfield made sure Gustav had no idea that his visitor was in the know. He smiled and joked in the Volvo back to the hotel, the file tucked down the back of his trousers. When he got back to the hotel, he put out the Do Not Disturb sign and locked the door. Then he heaved the large armchair in front it. Better safe than sorry, Bloomfield thought. This is national security, this could mean lives lost. Shutting the curtains, he turned on a bedside light, and pulled the file out of his trousers.

BY-PRODUCTS OF CREM
CLASSIFIED

As well as H3O, CREM's action causes the production of sulphydic nitrate, a highly toxic substance. Tests have shown that nine out of ten pigeons who breath in one mole of sulphydic nitrate are dead within twenty minutes. One CREM action has been found to produce ten moles per square centimetre of the substance, which, given its molecular weight, and depending on prevailing wind conditions, can spread as far as ten kilometres an hour.

Bloomfield stopped reading, his heart pounding. That Johansson was nothing but a filthy liar. The thing was a killer, it could wipe out birds, and what if cattle chewed on contaminated pigeons and then they were killed for beef and someone had a hamburger? It was Mad Bloody Cow all over again. He turned back to the document. In large black letters at the bottom of the page was the statement:

NO ALTERATIONS CAN BE MADE TO CREM TO AVOID THIS SIDE-EFFECT. WE RECOMMEND THE PROJECT BE IMMEDIATELY HALTED.

Bloomfield's palms were oozing sweat, but, at the same time, he was seized by the knowledge that someone had passed this on to him because they knew he could do something. He was the man for the job, he could take this to the right people, the higher-ups, and expose Johansson for what he really was. He could do it, and goddammit it, he would do it! Bloomfield put the file back into his trousers and started packing. His plane left in three hours. There was no time to lose.

The seatbelt signs lit up and Bloomfield leaned back in his seat (Economy. Best for travelling incognito). The old lady next to him was still rifling through her enormous handbag, and Bloomfield hoped that she'd stop that soon and then fall asleep for the whole duration of the flight, as old people tended to do. The hostesses, lovely girls, were coming round offering drinks. Bloomfield needed a stiff one. He could feel the file against his tailbone. He had taken it out and put it in his briefcase for the X-ray, in case they got suspicious (Sorry, sir, what is that paper down your trousers?), but had put it back when he got through to Duty Free.

Bloomfield looked out of the window. A light rain was falling. The aeroplane taxied along the runway, and then lifted up into the grey clouds. Bloomfield relaxed. He was off, not long now. He pressed the seat-back button.

Arriving at his fourth hotel room in as many days, Bloomfield set his briefcase down and extracted the file from its trouser hiding place. He uncrumpled the sheet and laid it on the desk. During the short plane ride, his mind had begun spinning. Nothing was clear any more. Perhaps he had overreacted? Perhaps this was a hoax, designed to test his mettle?

Bloomfield was exhausted. His fourth country in four days, he had a right to be a little frazzled. But he couldn't rest yet. He had to play his favourite game: Where's the Bible? He found it first go, in the top drawer in the left bedside table. Bloomfield punched the air triumphantly with his fist. He was the Master and Supreme Bible Locator! All these places were the same, he thought. Once you'd entered into their clever little mindset, you could crack 'em all.

He launched himself onto the bed, which sprang rather a little too much under his ample weight, which Bloomfield ascribed to unusually heavy bone structure. He lay back on the soft brown bed cover. A little snooze, that was what was needed.

Bloomfield slept heavily, dreaming of long, sweltering summers, of himself in shorts, bronzed as he had never been in his life (being prone to extreme sunburn), standing in front of his barbecue in his garden, adjacent not to the small (cosy) flat over the local supermarket but to the dream house he now shared with his (non-existent) beautiful wife and two (non-existent) beautiful kids. (Non-existent) beautiful friends milled around, the sun shone and shone and everyone was happy.

When Bloomfield woke up, rain was lashing at the window, and he knew what he had to do. He took the document into the gaudily tiled bathroom, and stood over the open toilet. As he shredded it he saw himself in his bedroom at home, his gorgeous wife kissing his deeply-browned chest as the children slept peacefully in their rooms. The scene became x-rated and Bloomfield sighed ecstatically. The world would thank him one day. Bloomfield closed his eyes and flushed.

THE ANGEL IN THE
CAR PARK

There is an angel sitting on the car park exit barrier. The rabbi can see it clearly, the outline of its wings, the glow of its goodness. Even from his car, three cars behind the empty vehicle at the head of the queue, he knows this is a heavenly being. He taps his fingers on the steering wheel and wonders what to do. Five minutes have passed since the driver of the car in front slammed his door and headed back towards the pay station. Five minutes in which the rabbi day-dreamed about his weekly sermon.

And now the angel.

The rabbi realises that the angel is looking straight at him. He thinks for a moment and then gets out of his car. How does one approach an angel? He inches past the cars in front, his heart pounding. All the while, the angel does not take its eyes off him.

He stands there, before the angel, and they look at one another. The rabbi cannot speak. He is struck as dumb as a baby. His feet have taken root. The angel sees into him, fingering his liver, his kidneys, his heart, touching every cell, every strand of his DNA. The angel is smiling, the angel is whispering, the angel is formulating something inside the rabbi.

The angel is gone.

The rabbi stands by the exit barrier, the life returning to his toes, his skin buzzing. He stands there and, for an instant, he knows all there is to know. It is all within him: the Tree of Knowledge, the Tree of Life, all of creation streaming alongside his blood vessels.

A loud hooting. A shout.

'Oi! Get out the way, mate!'

The rabbi turns. He gazes into the eyes of the man in the driver's seat of the car by the barrier, nods, and makes his way back to his Peugeot.

As he sits and waits, his sermon forms itself. When he gives it on Shabbat, his congregation will wonder where their rabbi has gone. Who is this man is who speaks to them with golden words? When they shake his hand after the service, they will walk away feeling as though they have touched something they have never known. And for the rest of that day, and perhaps the days that follow, they too will know something. On their way to work, sitting in a meeting, making dinner for the chidren, they will stop, remember, smile. They will be happy.

Then they will forget. Even the rabbi will begin to forget. The image of the angel will fade. And life will continue almost as it was before.

EVIE AND THE ARFIDS

Radio-frequency identification (RFID) tags are spreading, and they could soon be keeping tabs on every one of us.

—'Who's Keeping Tabs on your Tags?'
New Scientist, 28 August 2004

I get the job because I say I can keep a secret.

They say,

'Excellent, excellent, that's precisely what we need.'

They sit there, three of them, all in grey suits, one leg crossed over the other, all looking at each other and then looking at me and nodding. Like they were one person with three heads and six legs, all nodding at me at once. It's all the same to them that I've been doing nothing for the last thirty years, just raising kids, making breakfasts, lunches and dinners, cleaning, sorting, sweating, eating, losing my figure, losing my husband, watching my kids go off one by one, losing my mind.

'Mrs Applegate, you're a bright woman,' they say to me. 'No one has given you a chance in life. But we see your potential. We need a woman like you. You're exactly what we're looking for.'

I start on the Monday morning, and here's what I do all day: sit on my backside behind the machine, and watch stuff come out of a hole in the floor and along a belt towards me. The first week it's jumpers, with some logo on them, a fancy thing, Tommy Hilfinger, like the one I'd seen my Jimmy wearing once and he got all snippy when I asked where the money came from.

I press Pause, then push the green button and this arm comes out, grabs the collar and puts one of them little arfid things into the label. It sort of melts in there so no one can see.

Then I press Start, the jumper moves on and down the belt and through a hole in the wall and the next one comes. A half-blind monkey could've done it.

'It's secret work,' the Three Heads said. 'Classified. You understand. National security. You're doing important work, Mrs Applegate. Your country thanks you.' My country? Thanks, country, for the nice payslip, really nice, if you ask me. Way too much for just pushing a button and watching and not saying nothing to nobody.

The first day, I eat lunch by myself. At 12, I leave the machine on stand-by, like my video recorder when it's not really On or Off, and go out into the corridor. There are loads of doors but no noise. Maybe there are hundreds of other Evie Applegates pushing buttons just like me. Loads of us women who ain't got nothing better to do, who don't gossip, who ain't got no one at home to tell it to anyway. What makes me think I'm so special, I say to myself. I come round a corner and there's the canteen.

I get shepherd's pie and a bowl of fruit and look to see if there's anyone I can sit with. I can keep secrets but that don't mean I don't like polite chitter-chatter with someone while I eat. I'm not one of them who never speaks, like my Jack's friend Ronnie, sitting in our armchair and say nothing for hours and hours. I never understood why Jack bothered with him. Dumb Ronnie, I'd call him and Jack'd get angry.

'He's a good man, don't call him names, Evie,' he'd say at night while he tried to pull his trousers off standing up by the side of our bed. 'He don't say anything 'less he thinks it's worthwhile . . .'

'He must think I'm not worthwhile then,' I say, watching Jack struggle and holding myself in from shouting, 'For goodness' sake, just sit down and take them off, it's so much easier.'

'Don't start,' says Jack, 'with your "I'm just rubbish, no one listens to me".' And he gets his trousers off and goes to the loo. A month later he leaves me. I should have known.

There are only three other people in the canteen. It's quiet, in a spooky way. No one gives me a you're-new-here-sit-with-me look, so I don't. I sit by myself, eat the pie, which is OK, and the fruit salad. Takes me all of five minutes. What about my half hour? What do I do with the rest?

Nothing. I go back to work and push buttons until 5pm, then I put the machine back on Standby, get my coat on and go. I don't see no one then either. Strange. Aren't they all going home? I hope SanFirst International don't want me to work overtime.

I meet Gina on my second day. I go to the canteen and get stew and chocolate cake, which I shouldn't have but I'm feeling sorry for myself, thinking if I'm eating on my own then I'm going to have cake and it doesn't matter about my fat thighs, no one wants to look at me with no clothes on anyway. I'm heading for the same spot as the day before but then I see someone waving. A tiny thing with hair that's orange like a lollipop. I turn right around and head in her direction. No way I am eating alone if I don't have to.

'Hi,' she says. 'I'm Gina, nice to meet you.'

'Evie Applegate,' I say, and try to hold my tray in one hand and shake her hand with the other. She laughs.

'Here,' she says, and takes my tray.

'Thanks,' I say, sitting down. Gina's eating sandwiches. 'Oh, I didn't see any of them, do they make them for you specially?'

'I bring them from home,' she says, and laughs again like a little girl, but even though she's tiny, she's not young, she might even be coming up on my age, she just does it better.

We start chatting. She's really easy to talk to. I tell her about the kids, say quickly about Jack and me not being together. I want to ask how long she's been working here, what she does, but I'm supposed to be keeping secrets so

I don't. If she asks me, I'll say, I think, but only if she asks. Don't screw up this job, Evie.

At half past, she stands up and so I do too.

'Really lovely meeting you, Evie,' she says, and giggles. I giggle too, I couldn't not.

'I thought I was going to eat by my lonesome every day,' I say as we carry the trays back.

'It is rather quiet,' says Gina, bending down to slot her tray into the rack. 'Most people don't come to the canteen, they're too busy. But I'm in here every day. I need the break.'

'Glad to hear it,' I say, but then while I'm putting my tray away she says,

'Bye, Evie,' and walks off really fast. Bit rude. I guess she doesn't want me to know where she works. I walk slowly back to my room and get down to more button-pushing.

The second week, it's trousers. Armani and some others. A nice young man, Steve, comes and does some adjustments, so the arfid goes in the right place.

'We've got templates for most different kinds of clothes,' he says. 'And other . . .' He stops, goes all pink, and then turns back to the machine.

'Thanks,' I say.

'OK,' he says. 'Just do . . . whatever you did before.'

On the Monday of the third week, he's back.

'Shirts,' he says.

When they start coming down the belt, I have a good look but I've never heard of the name on these. I'm sure they're expensive too.

Lunch is the only thing I look forward to.

Gina and me, we get to talking about her.

'I don't believe it, you with two grown-up kids, and twins at that!' I say.

'I got married young,' Gina says, putting down her sandwich. 'Young and foolish, they say, and I certainly was.' She laughs a not-so-happy laugh and I know what that means right away. 'The children came along and it seemed like five minutes between them being in nappies and then off to university. It was a bit like a dream.'

'A dream with no sleep and lots of screaming,' I say. 'Not the kind of dream I thought it would be.'

'Actually,' says Gina, pushing her hair back so it doesn't get in her special smelly tea, 'the twins were very quiet. They were different from other people's children. I wondered why they didn't make much noise. I kept thinking, well, they're just getting used to being in this world, and when they're accustomed to it, they'll start expressing themselves. But they stayed quiet and that is just the way they are.' She doesn't laugh. She stares at me for a moment and then looks down into her tea. I put my hand on her arm.

'You can never tell how they're going to turn out, love,' I say. 'They're like little puppies that someone drops round your house one day, and you do what you can, you feed them and all, but in the end they're not you, they're someone else. You ain't never going to understand them, and they ain't never going to understand you. That's how it's meant to be.'

Gina smiles.

'Thank you, Evie,' she says. 'You're a good . . . listener.'

'Big ears, that's me!' I say. 'Just 'cause we're keeping secrets, don't mean we can't talk.'

When I say the s-word, Gina's face goes all funny, like the tea's gone down the wrong way. She looks at her tray, back at me, and then stands up.

'Oops,' she says. 'I'm late. Better run. Bye, see you.' And off she goes.

Damn, I think. Shouldn't have said that.

Then this week things start going loopy. On Tuesday, Gina comes into my room. I don't know how she knows it's my room, there isn't a hole in the door and all the doors look the same.

'Gina, what . . . ?' I say, standing up.

'Sit down,' she hisses, 'and just carry on, or they'll pile up.' I hurry to do the next arfid and out of the corner of my eye I see Gina pacing up and down.

'What is it, love?' I say.

She doesn't say a thing for about two-shirts'-worth of time. Then: 'I stole something.'

'Oh god, Gina, what did you take?'

She crouches down next to me.

'They're probably bugging us,' she says, 'but they probably know anyway, so it doesn't matter. I don't want to get you into trouble, but I had to tell someone.' It goes through my head that I could lose my job over this, but then my head says it doesn't matter.

'Tell me,' I say. 'What did you take?'

'I took a watch,' Gina whispers. 'With a tracking device in it. I needed it . . . and now I don't know what's going to happen.'

I want to put the stupid machine on Pause and give her a big hug and tell her it'll all be alright. She tells me the whole story, while I keep pushing buttons: a few weeks before I started, she'd come into this room and taken a fancy watch. Seems they put arfids in all sorts of things. The person who worked here before me got fired when the Three Heads found out that something didn't add up. Gina didn't say anything because she really needed it. With the arfid in it.

'I work in tracking,' she says. 'If you have one of those in something you're wearing, they can follow you everywhere. That's why I took one. I needed the money.'

'Money? Who were you going to sell it . . . ?' I stop. This isn't going to be something I should know about. But I don't say anything when Gina carries on:

'They approached me a year ago, offered me ten thousand pounds. They just wanted one. Ten thousand pounds! But I refused. I didn't want to compromise my job.' Gina makes a choking noise. 'Then Brian got into trouble, he lost a load of money, a friend of his, he'd lent it to for his new business, and the friend left the country, ran away with all of it and . . . and he called me, and he was crying and saying, "Mum, you've got to . . ."'

Right then, a voice comes over the tannoy.

Will all employees return to their stations please.

Until that minute, I didn't even know there was a tannoy. The place was always silent as a grave. It scares the wits out of me. Gina's as white as a bleached sheet.

'Shit,' she whispers, and the word sounds all wrong in her mouth. 'They know I'm in here. I have to go. Meet me at the bus stop when you finish.' I don't know what to say so I just nod and Gina slips out of the door.

I don't know how I get through the afternoon, all these things whizzing round my head: Gina stealing, her boy in trouble, ten thousand pounds, wouldn't I have done it for Jimmy? I do all the things I'm supposed to do, and then walk out a tiny bit faster than normal up the street to the bus stop. Gina's there, pale but smiling. She puts her arm through mine.

'Let's go get a cup of tea,' she says as the 147 pulls up.

Gina doesn't say anything until we have our teas in front of us and the waitress has gone. Then she shuffles her chair around so we're both in the corner with a wall behind us and no one nearby.

'I gave them the watch,' she says, 'and they paid me. Cash. I have never held so much money in my hands. I took it straight to Brian and gave it all to him. I told him that I had got a bonus and his grandmother had lent me the rest. He didn't ask. He was in such a state, the bank were going to take his flat . . .' She stopped and looked at me. 'I felt like this was the one time I could really be there for him. I had to do it. I had no choice. Do you see, Evie?'

'Of course, love,' I say, and put my arm around her. She's tiny, like a little girl. 'You're a great mum, a mother lion, doing whatever it takes for your kids. But what are you worried about? They fired someone, didn't they?'

'Something's not right,' she says slowly, 'I've been feeling it for a few days. I feel like they're watching what I do, and when I come in in the morning it looks the same but wrong. I can't tell you what exactly, but it makes my skin crawl.'

A shiver runs down my spine.

'What are you going to do?' I say.

'I already did it,' she says, looking proud and scared together. 'I put a tracker into me.'

'Oh my good god! What do you mean? Those things, they're not for that . . . they're for clothes and things, not . . . How did you . . . ?'

'I put my finger in the machine, modified it slightly, quick sting, and there it is.' She turns her hand over, and there's a little red square on the tip of the longest finger on her right hand. If you didn't know, you'd think it was hard skin. I'm gobsmacked. If you can put arfids into people . . . I remember what the Three Heads said about National Security. This is going too far.

'I need to ask you something, Evie,' Gina says, and when she says that, I have no idea what she wants. Does she want me to put one in myself too? Or in someone else? Before I can say, 'Hang on . . .' she says: 'If I am not in work, if something happens, I want you to track me.'

'Track you?' I say. I feel dizzy, like I might fall right off my chair.

'Yes,' Gina says. 'SanFirst have people . . . if they know what I've . . . I think that they might . . .' Gina starts crying.

'Oh love,' I say. 'Couldn't you go to the police?' But I already know the answer. They'd take the money away from Brian, and Gina won't have that, not for anything. 'OK,' I say. 'Tell me what I've gotta do—if something happens, but it isn't going to, don't you worry.' Gina looks up and tries to smile.

'Thank you,' she says . 'Evie, you don't know how much this means to me.'

For the next few days, Gina and I meet for lunch as normal and try to talk about ordinary things, like my Jeanie and her creep of a boyfriend, and some new food Gina's discovered, some strange seed. I keep sneaking a look at her finger. I just can't believe she's done that to herself. She's brave, if you ask me. Very brave.

Then, on Friday, she isn't there.

We'd arranged that she'd call if she was ill or there was some other reason. If she didn't call and she wasn't at lunch, I had to assume the worst, Gina said, and go with the plan. I sit with my tray in the canteen, my chicken drumsticks going cold, a sick feeling in my stomach, praying for Gina to just stroll up to me, a big smile, with her sandwiches and bad-smelling tea. I wait until the last minute, then put the tray back with all the food on it.

There's someone standing by my door. One of the Three Heads. This is it, I think.

'Mrs Applegate,' he says.

'Hi,' I said. 'Am I late? I was just getting back to . . .'

'Could you come with me for a moment?' He doesn't seem to be asking so I don't answer. We go down the

[66]

corridor in the opposite direction from the canteen. We get to one of the doors that looks like all the others, and he holds it open for me. The other two are sitting at a desk.

'Mrs Applegate,' says one. 'I am afraid we seem to have a situation here.'

My tummy is doing flip-flops.

'I believe you and Ms Workley have struck up an acquaintance,' says the middle head. They know, I think. Should I deny it? I don't know what to do, what would be best for Gina. I feel dizzy, but then I realise that I should just play it down. So we're friends, me and Gina. That's not a crime.

'Oh yes,' I say, in a jolly voice. 'She's my lunch companion, nice to have someone to talk to, you know.' I laugh a little. But they don't smile or nod.

'Ms Workley has been suspended.'

Suspended ain't the word, I think. Been disappeared is more like it.

'I'm sorry about that,' I say, keeping my voice all non-involved. 'She seemed like a nice person.'

'You understand that when something happens here, it doesn't just affect this company, Mrs Applegate,' the Three Heads say. 'This has ripples that go further than us. Much further. National security. You're a clever woman, Mrs Applegate, I am sure you appreciate the need for full candour in this matter. We are asking you to tell us what you know.' They all stare at me hard.

What I know, eh? I think. Clever woman, am I? So why aren't I doing something better than pushing buttons?

Why am I stuck in that room with a machine, doing the dirty work? I'm not clever, I'm stupid, I say to myself. I'm stupid to think that they would want me for anything important, for anything more than a little kid could do with just five minutes' practice. I've never done anything clever in my life. I let Jack leave me, when really I should have left him years before, should have taken the kids and gone. I would have been better on my own, without his sour face and his moaning and groaning. I could have done that, if I'd known which way was up, if I'd had an ounce of sense.

I look at the Three Heads, who are waiting for my answer.

'I'm sorry, but I don't know anything,' I say. 'Apart from what she eats for lunch, if that helps. She brings sandwiches, every day, makes them herself.' I sit back in my chair. Keeping secrets is what they took me on for, well, that's what they're going to get. I smile my sweetest smile and wait to see what they do next.

They don't do anything, of course. What can they do, since I'm not giving Gina up? They show me out and I walk back my room. The plan is in full swing now, and I'm going to do what I have to do.

At five o'clock exactly I leave the building. Gina gave me a mobile telephone and I keep taking it out of my pocket and praying she calls. But it never rings. Now is the Moment of Truth, and I feel like I did in one of those dreams I had when I was small, the night before a big test: I'm sitting in the classroom with everyone else writing and my mind all blank, and the panic just coming up inside me and everything spinning, and the teacher shouting at me, 'Eve Simpson! You are the stupidest girl we have ever had! Stand up! Everyone look at Eve Simpson!' and in my dream I stand up and they all laugh at me and point and throw paper airplanes.

'This isn't about you,' I say to myself. 'This is for Gina, and you don't have no choice but to do it.'

I pretend like I'm waiting for a bus, and then, when everyone gets on, I hide behind the bus stop, cross the road and walked back to SanFirst International. Around the side of the building, and there's the gate, like Gina told me, hidden among the bushes. There is a big plant pot, and I reach underneath and find the key Gina left me. She'd worked it all out.

The key's big and cold in my hand, which is shaking like I have Parkinson's or something. It takes me a couple of gos to get it in the lock, but finally I do, and I look around quickly before squeezing through the doorway. I'm in some kind of yard behind the building, with high walls on either side. There's a door straight ahead, with a little square of buttons with numbers on. I tell my body to stop its wobbling, take the piece of paper Gina gave me out of my bra, and type in the number. There's a click, and I push open the door.

Inside, it's quiet as a graveyard. I go down a corridor with only a small flickering light, and turn left and then right, and there it is: the Tracking room. Of course, it doesn't say *Tracking* or anything like that, just another white door, like all the others. Another number pad, I type in the number, there's a click and it's open. My tummy suddenly goes all queer. It's all going like we planned, and I think I'm going to be sick. I hold onto the doorpost for a minute, then stand up straight. Evie, she's counting on you, just do it, I say. I go inside and shut the door.

Gina told me that the room was always pretty dark, they didn't have any windows, just light from the computers. There are rows and rows of them, like they're all watching without anyone to watch them. Gina's computer is fourth one along on the second row on the left. I walk over, feeling like any minute all the computers are going to scream or something and the lights are going to go on and people are going to run in and grab me.

Nothing happens. Silent as my house on a Saturday night.

I sit, my handbag on the floor, the mobile phone by the keyboard. Then I wiggle the mouse thing backwards and forwards on the table. Even though Gina told me what would happen, I jump when the screen suddenly goes all bright. My heart is banging like a train.

On the screen are a lot of little pictures, and in the top right corner is a green tree. I take hold of the mouse again. I've used one of them things a few times, in the library, but it never felt natural. It takes me a while to move the arrow on the screen in the right direction, my hand gets all sweaty. Finally I get there, click the button and the whole screen changes. A line at the top says TrackID and I know I'm in the right place.

I get the piece of paper out of my bra again, and find the box on the screen to type in the number which means

Gina's arfid. It's very long, with letters and other things like question marks, and I get it wrong the first time. I'm sweating so much I worry I might flood the keyboard. I can't believe no one's heard me yet. It feels like I'm making loads of noise. I get the right number down and press the Enter button, and there's a big whirring. I grab the phone and my handbag and stand up, my heart going boom-boom-boom. But no one comes in and the whirring stops. I sit down again. Bloody hell, I say to myself, You're on the way to a heart attack if you keep this up.

The screen shows a map. At the top it says Roehampton Industrial Estate. Roehampton. Somewhere near Birmingham, I've never been there. There's a little red dot on the map.

Gina.

I touch the dot with my finger.

'It's ok, love,' I say. 'Someone's coming, we know where you are.' Around the map there are letters and numbers, and write them down. Then I pick up the mobile phone and follow Gina's instructions. I go into Messages, and then Write Message, and then I'm supposed to write, but all the little buttons have three letters on them. I start writing *Gina* but it comes out all funny, like *ggma*. I want to cry. What did I do wrong? Then by mistake I press the *g* twice and it turns into an *h* and suddenly I get it. I write: *Gina your mum in danger in Roehampton Industrial Estate 35D on map tell police from friend Evie.* It feels like it takes hours and my thumb is getting sore but I finish it. I press the Send button and start putting in Brian's number. Gina said it goes straight to his mobile phone. Seems like some kind of magic to me.

Just as I'm typing in the last number I hear something. It sounds like two people whispering outside the door. I panic, stand up, holding the phone and my handbag, and look around for somewhere to hide. There isn't anywhere just tables with computers on them and no windows, nowhere to go. My head's spinning round and round and I think I'm going to faint. The door opens.

'Mrs Applegate, put down the phone and put your hands up.'

Ohmigod, I think. Gina, what do I do! But Gina's just a red dot somewhere, a red dot in danger. It's just me now. They come into the room and they point guns at me and I'm breathing a thousand times a minute and all I can do is look at the phone and press the Send button and they run towards me but before they grab it I see it says Message Sent and I know I've done it. I sink down into the chair and put my hands up.

'Mrs Applegate, you are in serious trouble,' says the man holding the phone. The other man comes around and pulls me up out of the chair. He holds my hands behind my back and puts handcuffs on me, like you see on the telly. As the man pushes me by my shoulders towards the door, I'm suddenly very calm. 'Gina, love,' I tell her in my head, hoping she can hear me, 'it's all going to be alright.'

One of the men is talking into his phone, saying, 'We've got her but she made contact, I'm bringing her in.'

[72]

Bring me in, I think. Do your worst. You ain't getting nothing out of me. I smile a secret smile to myself as they drag me down the corridor.

FLORA COMES BACK

After he married that woman, Flora took up night jogging. She did violence to the pavement while others ate dinner, the air between street and sole of shoe the absorber of her shock. The imprints of his fingers remained, each whorl exactly where he last touched; a light dusting of dark powder and they would be revealed.

Back in her flat, she pulled off her trainers, and then waited as sweat slid from her forehead and stung her eyes. But still she couldn't cry. The ache in her jaw was insistent, a reminder her of how she slept alone, clenched. The thought of him and . . . Flora bent over, pain shooting across her abdomen, burning.

She took to her bed. Phone calls went unanswered. The paralysis of sleep cocooned her. She dreamed of journeys never taken, frozen Siberian railway stations, his warm kisses, tunnels wrapping them in blackness.

Then her dreams took her back to before men, to thirteen years old, to boys at the back of the class, girls giggling, Latin declensions, her father, his little girl, that woman who came into their lives, replacing something Flora never had, never wanted, never needed.

After a week, Flora emerged. She removed the laces from her running shoes, washed them in the sink and tied them to the ends of her plaited hair. She made coffee, and sat on the damp steps of her building. The air smelled of renewal. She still felt him, but softer now, and she sensed he would continue to soften and fade. She sipped coffee and watched the garbage men shooing away pigeons from an abandoned half-sandwich. She tipped her head back and watched the birds fly higher and higher, and finally, then, she cried.

RAINSTIFFNESS

Plenty of studies have looked at pain associated with weather, especially in people suffering from arthritis.

— 'Expect Pain in Knee Area'
New Scientist, 4 March 2006

When it rains, she stiffens. Lying in bed, just coming out of sleep, Billie smells the air, hears the tippy-tapping on her window, and knows that soon she won't be able to move. Deadness is settling into her jaw, neck, shoulders, chest, belly, legs, arms, fingers, toes. It's as if someone is covering her in caramel, soft and warm and slowly hardening until she can't even twitch. She can only blink. Blink and wait.

The first time, Billie was two, streams of water battering the roof. Her mother came into the room and, when Billie didn't begin her morning chattering, tried to wake her, shook her, even though the child's eyes were wide open, blinking. Her mother started screaming, her father came running, and he fetched the doctor, and they stood there in their terror as he examined her. The doctor searched and poked and investigated but couldn't find anything wrong. Finally, her mother pushed him aside and started stroking and squeezing the child, her arms, her legs, her whole body. The doctor and Billie's father watched, frozen.

Billie wriggled. Her toes, then her fingers, and then she waved her fat legs in the air. My baby, my baby! cried her mother, grabbing Billie to her. Sometimes, things happen to children when they are growing that we can't explain, said the doctor to Billie's father. They nodded and he left.

The second time, a few days later, for it was the beginning of a harsh winter, Billie's mother was less frightened. As grey clouds lined up across the sky and drizzle slid down the outside of the window, she sat by Billie's small bed and started rubbing the child's arms and legs. She rubbed and massaged and squeezed the sweet pink skin and Billie felt warmer and softer. After a while, she could sit up, move her head, talk again. Better now, she said and hopped down from the bed, stumbling a little.

As time went on, Billie's parents began to understand that what was happening to their daughter had something to do with the rain. They didn't know if the rain caused what happened to Billie, or if what brought on the rain brought with it this strange paralysis, but they knew that the two were intertwined. They knew Billie understood this too, although it was never spoken of. There's talk of storms tonight, love, Billie's father would say, and Billie would nod. She understood he was trying to prepare her, and she held herself back from saying It's going to happen anyway, Dad, nothing I can do about it. She knew that they tried to protect her from as much as they could in the world.

She didn't know that they had discussions in the deep of the night about taking her away, moving somewhere warm and dry and safe. Billie's mother would say, Where would we go? What would we do there? and Billie's father would reply, I don't know, love. It'd have to be somewhere far. Not sure I'd find work. And they would move closer to each other on their bed, and Billie's father would put his arm around his wife, and they would sit, unsaid thoughts in their heads.

These discussions went on, but even with both of them working, Billie's father as an engineer and her mother as a secretary, they barely had enough to live on. After a few years, they stopped talking about it, and that became the way things were.

They assumed they knew how Billie felt when it happened, assumed it must be terrifying. If they had asked, they would have been surprised to hear that Billie didn't mind it. It's raining, it's pouring, she sang to her dolls, My legs are all boring. I've turned into something hard and I can't get up in the morning. She dripped water down the outside of the dolls' house, picked them up and rubbed them all over until they were unstuck.

While it was actually happening, Billie lay quietly, listening to the rain. *I'm sliding, I'm slipping, I'm everywhere, I'm making things grow*, whispered the rain. *I'm cool, I'm wet, I'm small and fast*. And Billie replied, I'm a girl, I've got arms and legs, I can run and draw and play with my dolls. I can't do that now, I am in my sticky time, but Mummy will come soon and melt me out again. *Ah*, said the rain, *You're sticky, you're stuck, like my ice, you're frozen*. Yes, said Billie, Just like ice. Inside her she giggled because the rain knew.

When Billie was seventeen, she fell in love. Martin was two years' older. The night he asked her to marry him, on a walk through the churchyard, Billie put her finger to his lips. Wait, she said and ran home to her mother. What do I tell him? she cried, sitting at the kitchen table. Will he want me when he knows? Her mother, sleeves rolled up, kneading dough, pushed the hair from her face with one elbow. She looked into Billie's eyes and began pummelling the dough harder with her fists. She had known this time would come, but had prayed that it would not be so soon. Send him to me, she said.

Martin stood, fearful, in the doorway. Billie's mother forced a smile.

Come, she said, next to me, and he came and sat. Don't be frightened, she told him as she carried on working the dough. Billie has an affliction, it comes suddenly, but you can help her.

What? he said. What is it?

Billie's mother stopped kneading to look at him. The boy was pale as flour. She wanted to touch his cheek, but he was not her child. It's alright. You can make it alright. And she told him how.

When she had finished, he said, I can do that, I can do that!

I know you can, said Billie's mother. You must never . . . But Martin was already half way out of the door. She watched it close behind him, her heart tearing in two.

Now Billie and Martin have their own house, down the road from Billie's parents and around past the Post Office. It didn't rain on the June night they married, and not for three months after that, but Billie knows it is coming. She smells it in the air the night before. It's going to pour, she tells Martin sweetly. I will probably get stuck. Just touch me, rub me, don't be frightened, my love. But when he wakes up to ribbons of water streaming down the windows and his wife unmoving, staring at him and blinking, his heart races. What's ha—happened to you? I'll—I'll —get the doctor, I'll go—fetch him.

Billie blinks furiously and he looks into her face and sees that she is still there and that she needs him, and his heart slows down a little and he touches her cheek. Her skin is cool, and he can feel the muscles solid and unmoving. He starts stroking and massaging and moving down his wife's body and she begins to return to him and they keep looking at each other and looking at each other until she brings her hands to his face and pulls him down towards her.

When Billie finds out that she is pregnant, for the first time she feels fear. What if, when she stiffens, her baby becomes stuck? We have to leave, she screams at Martin. We have to go somewhere, anywhere else. He holds her hands, looks into her face and tries to know her terror. We'll go, he says. Tomorrow. It will be fine, my love. We'll find a place, far away. Just you and me and the baby. Billie and Martin pack up their things and Billie's parents drive them to the station. They watch as the train pulls away. She's never coming back, says her mother, resignation settling into her heart like ice. Come, says Billie's father, leading his wife away.

Billie and Martin travel across Europe as fast as the trains and their savings will get them. Billie watches grey-white clouds slip across the sky outside the train window, one hand stroking her swelling stomach, the other clutching her husband's. It's OK, it will be OK, Martin murmurs, but the look on her face tells him she doesn't believe him.

They travel down Italy to the foot, the heat of the carriage stifling them, their clothes sticking to their damp skin, the weight inside Billie growing heavier. From there they take a ferry to Greece, to a small island, and here they stop. They have run out of money and the strength to carry on.

For several months the skies are clear blue. Billie sits by the sea, watching and worried, while Martin works three jobs, in a restaurant, as a cleaner and as nightwatchman at the small port. When he rushes back home to change in the few minutes he has before his next shift, he searches his wife's face and, although she forces herself to smile and tell him she loves him, he cannot see her so clearly anymore. In spare moments at work, he sits with a cigarette and tries to remember the colour of her eyes and the sound of her laughter, but something inside him is growing cold and forgetful.

Then the winds change. They blow the painted shutters of their small house, and Billie's heart fails every time the shutters bang and then swing wide open. It's coming, it's coming, she whispers to her belly, to the child only a few weeks short of being born. It's coming, she screams to her husband and he tries to soothe her, but she will not be calmed. Finally, after several hours, she falls into a feverish sleep and he sits by her side, looking at her pale skin and at the mound of her underneath the bedsheets.

Outside, the wind beats at the house. Getting up to close the shutters, Martin stands, gazing at the dark clouds preparing their onslaught. He opens the window and the first drops of rain stroke his forehead and he smells freshness, renewal. He turns towards his sleeping wife. Her cheeks are pink and there is the smallest smile on her lips. He bends down and kisses her hand, its coolness startling his lips, and tucks it back under the sheet as the rain begins its battering.

Billie wakes as the downpour reaches its peak, drumming on the roof and knocking on the windows. She opens her eyes. No one is there. She blinks. She waits. She tries to sense what is going on inside her belly, but all she feels is the familiar stickiness spreading, setting. He will come, she thinks. He is outside making sure everything is shuttered, making sure we are watertight. He'll be back. Then she sends her thoughts to her baby. Don't worry my love. Whatever it is that you are feeling, I know it, it happens to me too, and we will get through this. Billie talks of her childhood, her mother and father, the village where she grew up. She tells her child stories, about playing with her friends, building make-believe worlds, wondering about what life would bring, falling in love.

Hours pass, but no one comes. Billie begins to panic. She tries to budge her arm, her leg, her neck as the deluge beats itself out onto the window. Nothing moves. She is set in stone.

Then she hears a voice. *Billie*, it calls. *I'm slipping, I'm sliding, I'm streaming, I'm coming. Breathe, Billie, breathe.* Billie smiles inside herself. He'll be here soon, Billie tells the rain. He'll come, he'll touch us, warm us, bring us back. And she closes her eyes and lets herself sink into the symphony of the storm.

HEART

She drew her hands out of the chest cavity and looked at the clock.

'Time of death,' she said.

In the locker room, she stripped off her bloodied scrubs and put on clothes for the real world. Then she left the hospital and turned the corner, rain flattening her hair.

At Sammy's, she sat at the bar, lit a cigarette and ordered a drink. When it came, she exhaled through her mouth, touched her fingertips to the rim of the glass, and remembered how it was to have a man's heart beat itself out in the cup of her palms.

KNOTTED

'Grew in their tummies,' I say. 'Nine months, swelled like a balloon. Then pop—and it's out.'

'Don't be daft,' she says, herself a Fully-Automated Reproduction: no eggs, no sperm, nought to eighteen in ten days.

'That's how it was,' I say, looking over her peachy skin, smooth blank teeth, fingernails shiny like they'd just taken the wrapping off.

'Urggh,' she wriggles. 'Tea?'

Doctors told me I could be one of the last, grow one inside me before everyone stopped doing it. 'No way,' I said. 'I'm for enjoying life, not creating it.' I wasn't wanting the wailing, the messed-up sleep, nipples chewed like Mars Bars. Had my head on the pillow eight hours a night, went out when I wanted, stayed in when I didn't, became one person and then another when I caught myself sticking.

Grabbed the world and didn't let go until my fingers creaked.

Peachy Skin comes back. She's pouring, but I see she's itching to ask.

'Did you . . . ?' she says. 'Have . . . one?' Nods to my tummy.

I chuckle but my old throat's been around the block, sounds more like growling.

'Nope.'

'You didn't want—not to be lonely?'

She's smarter than appearances, Miss Just-Unwrapped.

'Wasn't alone,' I say. 'Never unless I wanted.'

She comes around back and starts with the comb, trying to neaten me up. Then: 'At my other place, I like the children running around. Always something going on, someone to talk to.'

The comb sticks on a snarl of hair.

'Ow,' I say, eyes watering.

'Sorry,' she says, but a tear's already heading down my cheek. I wipe it away quick before Miss Perfect notices.

'Did I tell you about breast-feeding?' I say, closing my eyes as she fights with another knot.

EXPRESS

Speaking two languages slows the brain's decline with age, according to a study comparing monolingual and bilingual people.

— 'Juggling Languages Keeps Brain Sharper in Old Age'
New Scientist, 19th June 2004

You stroll through the terminal and take in the familiar other-ness. You smile. The regimented English, telling you where to walk, where to turn, where to leave trolleys; an unseen hand always guiding, pushing, reigning in stragglers. The country you have just left is looser, its fabric less tightly woven. Every Israeli paves his own road, shouts at the top of his voice, ignores instructions. A young country; even the old are still youthful in their rebelliousness. Here in England, you know what will not be tolerated. There is no raising of voices, no stepping out of line. You suck in the urge to disrupt, and instead gaze placidly at your fellow travellers. You wonder if they notice your sandals and cotton shirt, deliberately untucked and hanging over your loose trousers as if to say, I am not one of you.

You emerge from the Customs channel, declaring nothing, for all you have to declare is your return. It has been ten years, but you suspect the customs official won't care, won't welcome the returning son with cake or balloons; you would be lucky to get a smile. They don't care that your mother has summoned you to sit by her bedside and reconcile, and that coming back, for you, aches like a lost limb.

Pulling your suitcase, you walk past the waiting throng, knowing that no one waits for you, and you look around for a sign of how to proceed.

There it is. Heathrow Express. How many of your friends have sung the praises of this train? How you laughed at them.

'It's only a train,' you said, stirring your tea with its fresh mint leaves. 'So what?'

'It's a miracle,' they said. 'Straight into London, fifteen minutes. Straight in, just like that.'

'A miracle? Like the Red Sea?'

They looked at you as if at a small child.

'You'll see,' they said.

Now you will see. You pay what you feel is an extraordinary amount for the ticket, and make your way to the platform. Stepping onto the shiny train, you squeeze your case into the luggage rack and sit down. A television screen. Well, for what you have just paid, you should get some entertainment.

The television is showing a nature program. You watch with interest for a while, small furry creatures burrowing into the earth, lions and tigers chasing prey. The train fills up around you and you hear Italian, Spanish, French. Then—Hebrew. You turn your head, smile at the young woman talking into her cell phone. Someone is waiting for her. She will be there soon. Her words are home to you as you sit on this expensive fast train and smell England in the seat cushions, England in the window blinds, England in the businessmen with briefcases jumping on as the doors close. You shut your eyes and see the Jerusalem light reflecting off your bedroom wall.

You have to admit that the train is fast. You barely notice that fifteen minutes has gone and you are in Paddington. You locate the sign for the taxis and begin to cross the station. Underneath the large timetables with their flickering letters and numbers you stop.

The 12.15 to Bristol is now departing from platform 6, says the announcer. *Platform 6 for the 12.15 to Bristol.*

That is when you notice. The words pass through you without any effort, any thought or deliberation. They simply glide into your consciousness and are understood. For ten years you have listened to a language that was not your own. A decade of halting each word in its tracks to make sure that you have it right. At the beginning the process was slow and painful, one word sounding like five or fifty others, the context lost as you analysed and deciphered. Then it became more fluid, and you could swallow whole sentences without too much thinking. But the thinking was always there. Even after a decade, before making an important phone call you rehearse lines in your head. A phrase would catch on a branch in your brain and circle around and around until it hurt. The words were a song that you tried to learn with the right tune.

You tried to relax into this language as one would sink into a sofa. But a second language is not like a first; it comes after, and what was there before is always underneath, teasing you with its translations, tickling your mind, never just leaving you alone with your new vocabulary.

Sometimes, when you broke your teeth on a phrase you thought you had absorbed, you cursed and wished for a transplant, a chip that would turn you indigenous. You longed for a hard brush to scrub away your Englishness and leave you pure and fresh to start again, here, in your chosen home. But in Jerusalem you felt more English than ever.

You stand underneath the railway timetable and let people drift around you. It is as if time has slowed down for you only, and you remain in a daze, listening to the chatter, more announcements, a lost child, lovers crossing your path, a mother saying goodbye.

Will the parents of a small boy . . .

'I'll call you when I get in . . .'

The 12.25 to . . .

'Miss me!'

. . . has been delayed . . .

You drink in the words effortlessly, joyfully. All those years you thought you had become someone else, in the place you loved with all your heart. What you left behind was discarded, or so you believed. Now here, in Paddington Station, not two hours after landing, you understand that you are here to reclaim a part of you that deep inside you always knew was missing. You pick-up your case and start walking towards the taxi stand.

FIRSTS

Samantha watches him stroll, the folds of his trousers sliding smoothly against one another. She has never before seen a man absorb without arrogance the admiration of every woman. But then she has never before sat at a pavement table in a Venetian café in early summer. This is her first time in Italy, her first real espresso, her first self-possessed Italian man. Samantha sighs, the weight of all the firsts pressing upon her. Without it, she might fly.

Years later, when she tells her lover about that trip, she will laugh, raise her eyebrows, shake her head as if to dislodge the memories of that innocent time. Her lover will for an instant see her as she was - smooth skin, guileless eyes - and he will for a moment wonder how he would have loved her then.

EXCHANGE RATES

Your reproductive ability may be written in your hands. But what matters is the symmetry of your two hands and the length of your fingers, not the crinkles in your palms.

— 'Fertility Index'
New Scientist, 22 August 1998

It doesn't matter, does it. It has no bearing on the spinning of the earth on its axis, on the dollar exchange rate or the speed of light. Nothing will change because I have discovered that my uterus is useless, internal decoration, press a button and get an Error message, *Organs Out Of Service.* I haven't told anyone. I haven't told Rob. I just sit in my rocking chair ('so ideal for nursing'), staring at that damn blue sea print that he forced me to hang above the fake fireplace which is as defunct as my reproductive system. I am one of them, the Childless, the Barren, like Sarah in the Bible, but I wouldn't let my husband copulate with my handmaiden, even if I had one. The cleaning lady is Polish, she has several kids, she is rather unattractive. I don't think he would jump at that one. My mind is melting, I have no idea where the next train of thought will come from. I am going locomotive.

Rob went down on one knee.

'What are you doing down there?' I said. 'Wait a minute, what the hell are you doing?' I had lived fairly easily for thirty-one years with only a brief affair or two to disturb my equilibrium, in the certain knowledge that married was a state, like liquid or gas, that I would never become. I would remain solidly, stolidly, single.

'Marry me,' Rob said from the depths. 'Annie, marry me, for goodness' sake.'

'Why?' Not the cleverest response to a marriage proposal, but genuine. I liked him, but his proposal was the ice pick cracking at the berg of my beliefs. Everything I stood for was being challenged and I needed a reason.

'I love you. You love me.'

For the first time, I looked into a man's eyes and saw that he was telling me the truth. It was simple. In an instant I discarded my world view like an old sweater and began my metamorphosis.

Later on it occurred to me, as it never had before, that with wife-ness came the possibility of motherhood. Rob and I had not discussed the subject before our mad dash to the Registry Office, the image of white dresses, parents gushing and mushroom vol-au-vents making me nauseous. As we settled down to the day-to-day, the idea of a child tickled my mind lightly at first, then became a more insistent knocking. In bed, I said to Rob, 'Have you thought about us having . . . ?' I couldn't finish. It was unreal.

'Kids?'

'Yes. Kids. Children. Offspring. Me getting pregnant.' I was beginning to sweat. Rob held me tighter.

'Is that what you want?'

'I don't know. I just know that I need to begin thinking about it. What about you?'

'Honestly, I never really gave it a thought,' said Rob. 'I just wanted to be with you until we're wrinkled. I like kids, I never considered whether I wanted my own, our own.'

Over the weeks that came after, we mentioned it again and again, probing the space between us to see if it could contain new life. We slowly formulated a position: Use No Contraception and See What Happens. We did this, in many different ways. We enjoyed the doing, until six months of persistence resulted in zero return. My period was as regular as ever, never missing the chance to sign in red that month's death warrant of what had now become our great hope.

I said I would see someone. I began with my gynaecologist, moved on to a fertility specialist, and after months of probing, including donations from Rob, I had my answer.

In my dream I sit at an old wooden table and above my head swings a naked red light bulb. I am watching the light oscillate back and forth, back and forth, regular as the inside of a grandfather clock. No hand is pushing it but it doesn't stop. A woman walks in and sits opposite me, young, blonde and hostile.

'Why?' she shouts at me. 'Why aren't you doing it? What kind of a woman do you call yourself? Nothing's wrong with you, everything's fine, perfect working order, they told you, didn't they. Didn't they! It's you, you're doing this. Stop it. Stop!!' She is screaming at me, the light bulb swaying in a frenzy, red shadows spinning across my face. Then the red is blood gushing onto me, a miscarriage of the liquid of life and death, covering my clothes, making everything sticky. My blood smells of over-ripe strawberries. I wake up.

My mother used to tell me I was hard to give birth to. I fought it and the doctors had to pull me out.

'You were in a nice warm place, you wanted to stay, but I wanted you out,' she would say with a sigh. 'I was enormous, I felt like I filled every room I walked into, and I just wanted you out. I also wanted to meet you. But you didn't seem to want to meet me, or anyone.'

These conversations with my mother were never started by me. She was the one who brought up the subject, often, anxious to stress that I had always been difficult. As far as I could see, from my vantage point lower down on the family tree, she had provoked me at every turn. She was unsuited to motherhood, seeing everything I did or did not do as a direct insult. My father was silent. I vowed, as a child does, that I would not doom my progeny to a similar hell. However, I did not vow not to produce progeny. I just came to believe it was an option

I would never need to consider. Then, in an instant that the universe would not even bother to quantify in its wide concept of time, I received Rob, my hopes were awakened and then crushed like ice.

I rock for hours, the blue sea print approaching and receding.

The doctor had been extremely kind.

'I'm so sorry,' he said, sitting forward in his leather chair. I stared at the freckles on his large hands, clasped together on the desk. 'Your brain just isn't sending the correct signals to your ovaries, a kind of communication failure, if you will.' He smiled in a way that made me feel that he wished he could explain, that somehow knowing technically would help me feel better.

Communications failure. My insides: a faulty telephone exchange, with a plug pulled. Where was the operator?

'It's not uncommon,' he said. 'You're not alone.' But that was exactly how I felt. Sitting in the kindly specialist's office, I saw my fallopian tubes, my womb, my uterus, dissolving before my eyes, turning into powder which blew away in the breeze. My inside was a large hollow chest, locked. The key never even existed.

I don't will this, I think as I rock, my hands gripping the chair's arms. I don't will this, I don't want this, I can't do this. I stand suddenly, almost falling forwards as the chair keeps moving and hits the backs of my legs. Looking out of the window, it seems as if the streets have transformed into a parade of mothers and children, pregnant ladies with obscene swollen bellies laughing at my pain. The

[97]

world is a rolling sea of reproduction, in which everyone floats and I drown. Genes pass from generation to generation, life flows through and beyond, but here I stop. Everything I am and everything I am not, it all ends here. The red light never changes colour.

When Rob returns he finds me on the floor, in foetal position, the stance we all adopt in times of crisis. He comes to me, and together we sit and rock. You are not alone, says his body. You, mine replies, will never understand.

DRIZZLING

She sat down on the bench but the single page she had removed from the copy of *The Sun* she had found on the shelf between a tin of varnish and an old cardboard box wasn't enough; she still felt damp. She got up again and delicately tore several more pages from the newspaper. Just before she sat, she noticed that the top page was *that* one. The one with the naked . . . As she lowered herself down onto the newspaper, the thought that she was sitting on . . . sitting on *that* brought a small tingle to her knees.

Then she waited.

She listened to the cars driving too fast along their road. She was surprised she could hear them from this end of the garden. Faintly, but she could definitely hear. Richard and all his banging or whatever he does must drown it out. She noticed that it was suprisingly warm; she had expected a draft, a chill, but it was quite comfortable. She sat, her back stiff against the wall, and listened to the wind, the odd bird song, a cat wail. But what she was really listening for, what her ear was finely tuned to—footsteps, the handle turning—never came.

Eventually she looked at her watch, stood up, smoothed down her skirt, and opened the shed door.

Richard was in his armchair. He didn't look up when she came in. She stood for a moment and then went out again and into the kitchen. She took one mug from the cupboard, but as she turned on the kettle, she noticed there was newsprint on her hand. About to wash it off, she saw a word.

MEET, it said.

MEET.

She stood at the sink, gazing at the instruction embossed into her skin. The kettle clicked. She picked up the liquid soap and squeezed a little onto her hands.

Carrying two mugs into the living room, she set one down in front of Richard and sat on the sofa opposite. He raised his head.

'I . . .' he said. His face seemed crumpled.

'It's alright,' she said. 'It was our first try. Maybe next time.' He nodded, picked up the cup, and they sat together in the living room, drinking tea, while outside it began to drizzle.

PLAITS

Someone behind started plaiting my hair.

Hey! I said.

Sorry, he said. Just given up smoking, hands fidgety . . . hang on.

I sat there waiting for the lecture to start, feeling the gentle tug as he pulled one section of hair over the other. My knees said, Marry him. Don't turn round, just decide.

We married six months later. His face was as delicate as his hands were dextrous, his temper cool and his love eccentric. He washed my hair, made me pies with pastry messages on top, grew prize-winning roses, and said that the washing-up was his meditation.

When I cut my hair, he said that it was fine, that he could tickle my scalp now, but his hands were disappointed, and soon I smelled tobacco.

Stress at work, he said.

I talked to a wise friend, who said, Grow it back. My knees said, He should love you anyway. I saw him in a cafe with a woman I didn't know. His fingers were playing with her curls. I threw up in a rubbish bin and went home. I found a pack of his cigarettes and started one a day, even though my knees weren't pleased.

I love you, he said in bed, when my hair had reached my shoulders again.

I know, I said, and fought my knees' insistence that I go into the garden and dig up his rose bushes.

THE INCREDIBLE
EXPLODING
VICTOR

*Astronomers think white dwarf [stars] go through two phases
during their death throes. First they expand, and then they
explode.*

— 'Bubbles Trigger Explosive Death of White Dwarfs'
New Scientist, 22 May 2004

Victor Bloomfield was my best friend in junior school and
when he told me he was going to explode I believed him.

'It's gonna happen, it's in-evitable,' said Victor, taking
an enormous peanut butter sandwich out of his
Superman lunchbox. He bit into it, chewed for a while,
and then said,'It's not so bad, I don't think it'll hurt.' He
shuffled around to face me. 'Howie, probably best not to
stand too close when I feel it coming. It's going to be
messy.'

'OK,' I said.

Victor's explosion was pretty much all we talked about. His mother fed him too much when he was little and so his body grew more than it was supposed to, he said, and now he had to eat to keep up with what she had started. One day his body would give up, and Victor would come apart with ginormous energy, all his bits waving goodbye to each other and whistling through the universe.

'It's like when a star dies,' he said as we lay on the grass at the front of my house in West Hampstead, looking up at the sky, fanning ourselves with comics and sweating. It was summer and London was one gigantic oven. 'A super-nova. A big bang, and all that dust and stuff, and the star gets spread across the whole sky, in tiny bits, and some of those bits get attracted to other stars and become new stars. That's what's gonna happen.'

He made it sound beautiful.

Victor was an only child. His Dad died of a heart attack when he was two and so it was just him and his mother. Mrs Bloomfield was a really big woman. She wore tight clothes which you couldn't look at for too long without wondering how the buttons didn't just pop off or the seams rip apart. She wobbled and waddled around her kitchen, feeding anything that came in. Feeding was the only thing Mrs Bloomfield liked to do. She was from Germany and she had been through stuff in the war and she never wanted anyone to be without food, Victor told me the first time he invited me over. He said I had to eat something. I did eat, the first few times. But after listening to Victor, I decided that I wasn't going to let her stuff me the way she stuffed him, even though I was really skinny and never really put on any weight. But I couldn't risk it.

'Oh, no thank you, Mrs B, I already ate,' I said.

Her face didn't just fall, it sort of collapsed. The colour went from her huge cheeks, she folded her hands across her stomach and then went around the kitchen straightening pots and pans and muttering things we couldn't hear.

Victor, his mouth full, stared at me across the table with a look that was a mix of being really jealous that I was brave enough to say no to food and worrying like he always did that this might be his last mouthful before the end. I just sat there and watched as Mrs B focused all her attention on Victor, bring out plate after plate: soups, meat, fish, cake, biscuits. Couldn't she see what she was doing? How could she not know that she was turning Victor into a walking time-bomb?

When your best friend is convinced that the forces holding everything inside his skin aren't going to be able to do it much longer, you get caught up in all of it. I mean, you want to tell him that no way is he going to explode, but there's a little teeny corner in your mind that's thinking, What if . . . ? So I tried to find out everything I could about people exploding.

'What a crock of shit,' said my brother who was thirteen and knew almost everything about everything, as far as I was concerned. He stared at me. 'Victor's a fat kid, and he's probably going to swell up like a pumpkin if he keeps on eating the way he does, but he's not going to explode. People don't just blow up. God, Howie, how dumb are you.' And he pushed me out of his bedroom and shut the door.

I wanted to say something to my dad, but didn't want to tell him exactly who I was talking about in case he thought I was nuts and sent me to the kind of doctor that Simon Garfield had to go to when he told his mum he liked eating grass.

'Dad,' I said one night after dinner when we were loading the dishwasher that we had only just got and that still held enough excitement for me to offer to help out.

'Hold that,' he said, giving me a handful of dirty knives and forks while he fiddled with the cutlery holder.

'Dad, have you ever heard of someone exploding?'

He took the cutlery from me, bent down and put it all in without sorting out the knives from the forks, which I knew would make Mum swear when she emptied it. I waited.

'You know,' he said as he stood up, 'there is something called spontaneous combustion. I've heard of it, when a person just spontaneously explodes.'

'Really?' Blimey, maybe Victor wasn't so crazy after all.

I looked up 'spontaneous combustion' in the enormous old Oxford dictionary that Dad had from when he was actually at Oxford, forty million years ago. The book nearly crushed my knees as I sat on the floor and turned the thin pages really slowly in case they fell apart. There wasn't anything under 'spontaneous combustion', but there was loads under 'combustion', about chemical reactions and things catching fire. I started wondering about Victor catching fire. Maybe all that food would turn into something dangerous that leaked out through his skin and if someone had a match nearby he would just go off like a firework. I sat there with the dictionary in my lap, thinking of Victor the Firework shooting into the sky followed by a trail of sparks. Victor would like that, I thought. What a way to go.

But then I'd be alone. I didn't have any other friends, no one else at school talked to me, so life would be pretty miserable. I couldn't let that happen.

'Come on, you can do it, Victor, it's a matter of life and death,' I said, doing my best Captain Kirk imitation.

'I can't, Howie,' wailed Victor, looking like he was going to cry. 'I can't lose weight, my mum won't let me. She says I'm just right and if I don't eat everything she feeds me, I'll waste away. I can't not eat all the food. How? I can't!' Victor was fiddling with his hair, twisting it around his fingers, his cheeks pink.

'Look,' I said, trying to think logically like Spock. 'OK, let's just examine it again. Your mum makes you eat. You don't want to. You have to do something with the food so she thinks you ate it, but you didn't. And then, when you lose weight you just keep on wearing baggy clothes, and maybe a cushion or something, and she'll never know.'

'But how do I hide the food?'

I thought for a while. Victor chewed his thumbnail.

'OK,' I said. 'I'll get you a plastic bag from the kitchen, and you hide it when you sit down for dinner. Then, whenever she's not looking, you stuff food into it.'

'But she watches me eat.'

'Then ask her to get you something, a drink or something, or make her go out of the room.'

'Every time?'

'Yes,' I said, feeling like we'd hit on the winning plan. 'Every time.'

It worked, at first. Victor came up with different ways to get her out of the room. First he asked her to get him his jumper, saying it was too cold in the kitchen even though his mother was sweating from the heat of the oven; the next day he said he thought he heard someone at the door and she waddled off to see who it could be. While she was gone, Victor quickly stuffed a couple of potatoes and half a pie in his plastic bag, shoved it up his shirt, and stuffed his mouth with the rest of the food as if he'd eaten it all.

After two days of this, when he said could she get his school books so he could show her something he had done in class, she said no.

'I stay here. We will eat together,' she said, and she sat down beside him, and watched.

'Damn,' I said, when Victor told me. Well, if she wouldn't go when he asked her to, we had to find another way. 'OK, I've got an idea.' I started calling his house at exactly the time in knew Mrs Bloomfield, who was pretty clockwork about mealtimes, was giving Victor his dinner. Victor said she got all flustered when the phone rang, and dashed out. I hung up, but it was enough time for him to stuff a couple of potatoes in his plastic bag. Except that if we had thought about it properly we would have realised that this tactic wasn't going to last. After three days of her rushing to get the phone and it being hung up at exactly the same time every day, Mrs B got suspicious.

'This is one of your friends, yes? What games are they playing!'

Victor couldn't lie to her face. He pulled out the plastic bag. He told me afterwards that his mother went white, then green, then red. She grabbed the bag, screamed something at him in German, said something about concentration camps, and didn't speak to him for the whole rest of the evening.

The next day, mealtime was back to normal: Mrs B dished up and Victor ate.

'I'm gonna explode, Howie, there's nothing for it,' he said sadly the day after our failure, as we walked up the hill from Finchley Road towards school. It was pretty steep and I was walking slowly, as usual, so that Victor didn't have to huff and puff too much. Sometimes when I looked at him I thought of a big wounded animal like on David Attenborough, maybe an elephant, dragging itself around, knowing that it's not going to make it much longer.

'OK,' I said. 'Well, if she's going to make you eat all the food, then we have to make sure it doesn't stay inside you.'

'What do you mean?' said Victor, stopping to lean against a low wall and catch his breath.

'You could puke.'

Victor looked shocked.

'I can't do that, that's disgusting, that's just yuk. No, don't make me do that, I can't, I just can't.'

'OK, alright, well maybe you could go to the loo more.'

'Oh!' Victor's big face brightened. 'I can do that.'

I knew that my mum had these tablets that she took when she couldn't go to the loo, like for days. It happened every couple of months, she said to my dad it was like having a plug up her bottom, but I don't think I was supposed to hear that. They were called Lax-something-or-other, and I knew where they were.

Victor took two after dinner, and reported to me that it worked 'really really quickly'. He tiptoed to the loo and tried not to use too much loo paper otherwise his mother would have been suspicious.

'And the best thing, Howie,' said Victor, almost singing when I saw him the next day, 'is that I felt un-full. I haven't felt not full for so long. It was amazing. Like I hadn't eaten at all.'

We thought we had found the answer: Mrs B would feed him and Victor would take the pills and it would all go straight through and come out the other end. It was a perfect plan. I was very proud of myself.

After a week, Victor got really excited because his trousers were a teeny bit looser. He was starting to compare himself to Clark Kent and imagine a time when he didn't wobble when he walked and when he could actually run when we played football. But he was going to the loo all the time, and he was starting to feel sick. It made him sore 'down there,' he said, and when we walked uphill to school, he felt kind of dizzy and had to stop quite often.

'Howie, is this normal, do you think?' he said to me as we sat on a wall on the way up the hill.

'I don't know, I don't think my mum ever took them for this long,' I said. I didn't want to think that my plan might have any problems, but I was also getting a bit worried about what we were doing. When I got home I sneaked upstairs to Mum's bathroom and read the label on the bottle. Then I ran into Mum and Dad's room while they were arguing over whose turn it was to load the dishwasher and called Victor.

'Stop taking them!' I said. 'It says you shouldn't take them for more than two days.'

'Howie!' wailed Victor. 'Am I going to die? I took loads, what's going to happen?'

'It'll be OK, just don't take anymore.' I hung up and sat on the bed. What if I'd almost killed my best friend? I was only trying to save him.

The next day, when Victor didn't meet me on the corner I started to panic. I waited until I couldn't wait any longer and then ran all the way to school. I thought maybe he would come later, maybe he was just tired. But he didn't come all day and my heart beat really fast all through Double Chemistry. I've killed him, I've killed Victor. He's dead, and I just wanted him to get thin and I screwed up everything and he died and it's all my fault and they'll put me in prison and Mum and Dad are going to be really angry.

It took all my courage to knock on Victor's front door.

'Howard,' said Mrs Bloomfield. She looked very stern. 'Victor cannot play today. He is unwell. He is in bed.' I tried to see past her but she filled the whole doorway like a giant mother whale.

'What's wrong?' I said in a very small and polite voice.

'He is having a bad stomach.'

'Can I see him? I'll be very quiet, I promise.'

'Very quiet, yes? He is not to be excited.' She moved slowly to one side and I inched past. She nodded to the stairs.

Victor was lying on loads of pillows, his knees up, the whole bed covered in comics. He didn't look dead, his face was pinkish and he was definitely breathing. I let out the breath I had been holding all the way up the stairs.

'Hey, Howie,' he said, and his voice wasn't normal Victor voice but it wasn't so bad. He grinned. 'Mum got me the latest one, look!' And he held up the Superman comic we'd been saving up for.

'Cool,' I said, and moved nearer to the bed. 'Are you OK?'

'The doctor came,' said Victor, his face screwing up. 'My tummy really hurt and Mum was really worried because I was on the loo all the time. He pressed everywhere and then he made Mum leave the room. And I told him.'

My heart dropped into my shoes. It was all over.

'What did he say?' I whispered.

'He said we were really really stupid and it could have been really bad and dangerous, but luckily I stopped in time and I'll be OK, but it'll take a few days.'

'Oh god,' I said, and kind of wanted to cry a little, but I didn't.

'He asked me why I did it, and I told him about Mum and the food that she makes me eat and how I can't not eat it.'

'What did he say?'

Victor leaned towards me looking very serious.

'He said I had to understand that my mother thinks that she is doing the best thing for me, giving me lots of food, that's her way of saying she loves me. And he said that he would talk to her and try and make her see that it's just a bit too much.'

'Really?'

'He was great,' said Victor, leaning back against his pillows. He grinned. 'I think it's going to be OK. Mum came and asked if I was hungry, and I said No. And she didn't say anything.'

'Wow,' I said.

'Yeah,' said Victor. 'I think she gets it. Sometimes mothers just don't get it until someone like a doctor says something, you know?'

'Definitely,' I said, feeling very happy that Victor wasn't going to die and maybe not explode either. 'Has that one got "The Atom" in?' I said, looking at the comic.

'It's so great, you'll never believe what he does,' said Victor. 'It's called *Seven Foot Two and Still Growing*,' and he shuffled up and bit so I could sit on the bed with him.

'Go on,' I said. 'Read it.' Victor turned to the first page where a giant Superman had grown bigger than a skyscraper.

'*Lex Luther*,' says The Atom, '*Hahaha, I've finally come up with a sure-fire way to defeat Superman*,' read Victor dramatically, and we grinned at each other. Victor put on his best Superman voice: '*I'm growing out of control, my Super-Body is running wild!*'

Downstairs I heard Mrs Bloomfield switch on the radio and the sound of some German opera singer floated up the stairs.

'We'll do a page each,' said Victor, and passed me the comic.

YOU'LL KNOW

Prospective parents who are unmarried, over 50 or obese will not be able to adopt children from China under new rules, US adoption agencies say.

—BBC News, 20 Dec 2006

I agreed because I wanted you so much.

'Come on, love, you can do it,' Bill said. 'What's a couple of pounds? Half a stone?'

'Fat woman no get baby,' said the official, and her English and my Mandarin weren't good enough to get into an argument.

'OK,' I said, and started dieting.

We came back when I managed to get down to ten stone. It took two months, a personal trainer and constant reminders of you. Every time I wanted chocolate, every time I had to have that second helping of potatoes, I got out that picture of you and looked into your big black eyes and that did it for me. If I had to be thin, I'd be thin. Well, thinner.

This time there was a different official.

'I did it!' I said. 'Lost the weight, look!' and I spun round like I was on one of those shows where before they're hideously blobby and by the end they look normal but their skin's better and someone did their make-up.

'New regulation,' she said, not even examing my waist size. 'You take baby but you give us some thing.'

'I can lose more, if you need me to,' I said, and Bill gripped my hand tightly. 'Or money, we have more money.'

'No money, no lose weight,' she said. When she said what they did want, I felt sick. Bill was green. We went into a corner to whisper.

'No way,' he said. 'Are you crazy? These people are lunatics. Let's go. Let's just go.'

'I can't leave without her,' I told him. 'I'm sorry, but if that's what it takes . . .' Bill gave me a look but he didn't say anything else. He just put his arm around me.

The doctor was very kind. He didn't speak any English but behind his eyes was a gentleness. The nurse gave me the injection, and Bill was there, holding me hand, looking at me with such love and fear, but I never wavered, not for a second. The whole thing took an hour, and then when I was just waking up, the official came in. With her was the social worker. And with her was you. You lay in the social worker's arms, and I fell in love all over again. Bill helped me sit up, they put you in my lap, and it's like you'd always been there.

When we brought you home, things were hard, of course. You cried a lot and we didn't know why. But when you smiled, we melted. Bill even more than I did, I think. He hadn't expected this; he thought he was doing it for me.

'I can't imagine life without her,' he said a month afterwards. 'You were right.'

'All worth it,' I said.

Of course, it's difficult doing your nappy, and picking up your bottle, things like that. But I'm getting used to it. I find ways, it's amazing how you compensate. Bill says he's sure there are prosthetics, we can find someone who'll do it. 'We'll get you new ones that'll work even better, you'll be Bionic Mummy,' he says, but you know what? I don't mind it. Why? Because when you grow up, you'll see the gap, I'll tell you the story, and you'll know. You'll always know when you look at my hands just how much I love you. There's nothing I wouldn't do for you. Nothing.

MY NAME IS HENRY

'My name is Henry. You can't disagree with that.'
'No, Henry, I don't disagree.'
'Good, that's good. Henry. My name is Henry.'
'Henry, do you know where you are?'
'Where I am. Of course. Where am I? Silly question.'
'Where are you? Henry, where are we?'
Silence.
'Henry. Where are we?'

JANUARY 2ND: 10 A.M.

The young man paces up and down the small room
with one bed and a single high window.

'Henry, my name is Henry,' he mutters. 'Henry is my
name. That's right. Henry. They can't take that away, can
they? No they can't.' Inside his head, pictures come and

go, and his mind tries to catch hold of one and make it stay, but nothing remains. 'A tree, maybe, could be a tree. Grass. Or the sea,' he tells himself. He cannot stop moving in this limited space, he walks up and down the length of it, up and down, as if he is on a long road that leads out of a city, a road with no end.

The man is young because his life has not yet begun. He is only just stepping into adulthood with its responsibilities and desires, joy and insecurity, irrational love and loneliness. The doctor who is watching him on her video screen in the next room sighs. She sees the waste of a perfectly good human being every day, and she knows well that the chances of his recovery are not great. She sighs because this is the profession she has chosen and the place she is in. Her friends and her lovers cannot understand how the doctor can endure this work with the damaged and the confused, but she does. It replaces something from her childhood that she does not even realize was missing.

JANUARY 1ST: 8 P.M.

'No, no, let go of me!'

'His name is Henry Hunt,' says the orderly.

'Henry,' says the doctor gently. 'Henry, there is no need to be afraid. We are here to help.'

'Let go of me! Let go!' and the young man lashes out at the orderly, his arms swinging wildly. The orderly, a big man experienced in these situations, catches hold of the young man's arms and, without inflicting any pain, holds him so that he cannot move. The young man is suddenly quiet, subdued. He does not look around him, he looks only straight ahead, at the wall.

The doctor feels a surprising sympathetic pain for this young man, hardly more than a boy, who has come here and does not know why, who left his life behind only a few hours ago, and is now somewhere that he does not recognise. His face is like an angel, thinks the doctor, and is amazed at the thought. She believes herself to be detached from the patients in her care, and this thought intrudes on her distance. Yet she cannot help but look into his face, the blue eyes that she imagines were once bright and are now faded like aged paint, the red lips that may hardly have been kissed, the soft skin of his cheeks with a faintness of stubble. The doctor tries to contain an emotion rising in her. 'Henry, Bruce will take you to your room,' says the doctor and nods to the orderly. When they have gone through the double security doors, the doctor walks slowly to her office, sits in her chair and puts her head down on her desk.

JANUARY 1ST: 4 P.M.

'Can I help you?'

'Henry, my name is Henry,' mutters the young man, standing in the middle of the lawn. He is wearing a damp suede jacket, and his hair is wild around his face.

'Can I help you, love?' says the woman more kindly. She sees that the young man is 'not all there', as she will later tell her husband and their two children, 'standing there, on my grass, looking around him, like he's in a foreign country,' she will say, excited that, unlike her everyday routine of cleaning, ironing and reading magazines, today she has a tale to tell to her family, who listen as they spoon mashed potatoes and beans into their mouths. 'Didn't take me long to realise, he's not all there, poor chap,' she will say. 'Looking all ragged, like he'd slept under a hedge.'

'Are you alright, love?' she says to the young man and walks towards him, not too fast in case it scares him.

'Henry, my name,' mutters the young man. He bends down and runs his fingers through the grass. 'Green,' he says. 'Blue, red, pink.'

'It's grass, love,' says the woman, who has stopped a few feet away, because he is young and strong and if he's not quite right in the head, he could get violent, she thinks, although she likes the way he is feeling her lawn, a sensitive chap. But when the young man drops onto the grass and lies down on his stomach, the woman is afraid. What if he doesn't move, she thinks. She hurries into the house, making sure to close the front door behind her. She picks up the phone and for a moment does not know who to call. So she calls her husband.

'Call the police, Doreen,' he says. 'Quickly, before he smashes a window or something.' She puts the phone down, her hands shaking now with the horror that could be, and she picks it up again to dial 999.

'There's a man,' she tells them. 'Not right in the head, he's on my grass.' The police are coming, they tell her, and she moves to the front window from where she watches the young man, who is still lying there, now on his back, staring up at the sky.

JANUARY 1ST: 3 A.M.

The young man is crossing the park, rain soaking through his thin suede jacket. He is on his way home from a party celebrating the New Year, and all that is in his head is one word: Katy. Her smile, her eyes, her touch on his arm as they sat together on the sofa. Katy. The young man runs faster, his arms out wide, ignoring the storm. He does not even feel the cold and wet. Something else is flowing through his veins, the first hints of a first love.

The young man suddenly stops and takes his mobile phone out of his pocket. He presses the Address Book button and the screen lights up. He enters *K, A*. The telephone recognises what he is doing, and *Katy* flashes onto the screen. The young man grins, the phone number she gave him is stored there safely and tomorrow he will use it. Now that they have spoken, now he knows that she is aware of his existence, he can call her. He will call her. He has fluttering birds in his stomach. *Hi Katy, it's Henry*, he rehearses in his head. *Katy? Hi, it's Henry.*

The thousands of volts that hit him at that moment are not registered in his conscious mind. He had not noticed the thunder and the fierce wind shaking the trees around him; he did not notice anything. The electricity comes down from the sky so fast that there is no time for him to look up and see it arrive. It reaches into the corners of his brain and the heat it creates is something human neurons are not designed to withstand. The fingers of electricity seek out different parts of the young man's body, searching for a way through to the earth. When they find ground, the young man's knees buckle, and he slides down onto the sodden grass. All around him, the storm continues, while the young man's brain struggles to piece itself back together.

DECEMBER 20TH: 7 P.M.

The young man lets the elderly woman with her shopping get on the bus first and then he goes up to the top level and sits at the front. The bus starts with a jerk and then moves slowly down the street. The young man is thinking about the History lecture that has just ended. He is only in his first term at university, and is still unsure about what he is doing and where he is leading himself in life. History and Economics seemed a fine title to a degree, one that the young man's parents approved of. Yet he was drawn to names that spoke of different worlds: Psychology, Philosophy. Perhaps later, he will have his chance to delve into other realms.

Today, though, he is thinking about History not because of the wars they are analysing, but because of the girl he was watching. He has watched her at every lecture since the first, in September. He turns to look at her when the lecturer asks a question, for she invariably raises her hand. Her answers are always what the lecturer wants to hear; the lecturer smiles while she is speaking as if thinking, Why can't they all be this excited about what I'm teaching them? The young man is excited, but not about the question or the answer. He is excited about the girl, whose name is Katy. She shines so brightly in the lecture hall that he does not understand why everyone is not dazzled by her. He has tried to move closer to her, and with every lecture he is a few seats nearer, but she has friends who surround her and he cannot breach their stronghold.

The young man leans forward, his elbows on his knees, his chin in his hands, and stares out at the road in front of him, but sees Katy's face. He has always been too shy to speak to girls. I will talk to her, he thinks. A fear spreads through him at the thought, but at the same time, he is already impressed by his own braveness. A change is coming over him. He thinks of a film he saw where the hero and the heroine keep missing each other, one walking into a room seconds after the other has left. He laughed while he was watching, but a part of him knew that feeling. Next week, I'll say something to her, he thinks, as he stands up, rings the bell and makes his way down the aisle to the stairs.

When lightning happens, electricity flows between sky and ground, packing a wallop of 10,000 to 200,000 amps . . . Victims often have several discrete areas of damage dotted around the brain— dubbed the 'Swiss Cheese Effect'

— 'Aftershocks—Living Through a Lightning Strike'
New Scientist, 25th August 2005

FISH-FILLED SEA

She hears the shower running and she stands outside the door until he emerges, tea tree and lemon on his skin, flowers and fruit conditioning his hair, and she presses her hands over her mouth to stop herself retching as he bends to kiss her goodbye, a look of patient resignation on his face. He whistles down the stairs, the door bangs, and she stumbles into the kitchen, opens the rubbish bin and inhales. Her dizziness begins to subside as the evil stench is diluted by the old tea bags, coffee grinds, plastic wrappers with remnants of ancient sandwiches.

She feels better.

She gets dressed, switches on her computer, and begins her workday.

At six she switches it off again and waits for him. This is her time. If the morning is confrontation, and she loses daily, the evening is victory. He has submitted to her arguments about the wastefulness of two showers a day; she appealed to his sense of energy injustice with pleas for water conservation, and to his love of animals with tales of horror caused by synthetic confections making their way down the pipes and out towards the fish-filled sea.

Fine, he wouldn't bathe at night. She beamed.

She sits on the sofa, drinking tea, straining for the sound of his car returning. The key turns in the lock and she is up and towards him, grabs him by the arms, pulls him to her, gulps in his scent, and he grins. She wrestles him out of his jacket and drags him up the stairs, clothes peeling off as they go. On the bed, she pushes him down and slides along his body until her face reaches the sacred trough where arm and body meet, and she buries herself inside it. There, no trace of tea tree and lemon has survived; there, all the day's bodily products thrive, greeting her, the musk of his exertions, and she sniffs long and deep, drinking him in. He lies back, looking up at the ceiling, feeling her warm breath against his sticky skin, and he stretches his other arm towards her, stroking her hair and listening to the small snuffling sounds coming from his armpit.

NORTH COLD

Global warming can be intangible, but melting ice: that we can see.

— 'Global Warming—the Flaw in the Thaw'
New Scientist, 27th August 2005

There is a small town in the north that is cold all year round. The people living in the town have come to accept it as they accept that the sun sets and the sun rises. Cold is the natural order of things. No one talks about the weather except visitors, who do not stay for longer than they must. The people of this town have thicker skin than those living elsewhere; here, people live further inside themselves.

The people of this town have fashioned their own measures against the cold, although *against* is not the correct word, for they do not view it as an enemy. They never leave their houses without several layers of clothing, the outermost being woollen. They wear a hat to bed and sleep under many layers of blankets. You will not be surprised to learn that not many babies are born in this town. Cold is effective at dampening even the most burning ardour.

So, it was that on a Thursday in January a young man arrived. He descended from the bus, which made its journey from the south once daily, and rubbed his hands together.

'My, my,' he said to no one in particular, 'they were certainly right about this,' and he blew on his hands, whose bareness was already attracting stares from the people passing. No one said anything. What should they say? 'Wrap up warm, son, you'll catch your death?' The people of the town were not so forthright. They preferred to be left to their business and not to poke into the business of others. *What comes will come, to one and all*, could have been the town's motto.

The young man looked around him with a great curiosity. I am here, his wide eyes said to the townspeople. He picked up his large bag from the frozen ground, slung it over his shoulder, and began to walk with purpose into the centre of the town as if it were not completely foreign to him. The young man, whose name was Sergeant, possessed a confidence that everything he needed he would find, and so he strode on, never thinking of asking for directions to the hotel.

This being a small town, the hotel was where it was expected to be: on the main street.

'I would like a room,' Sergeant said to the man at the desk, setting his bag down beside him.

The older man looked at the younger, who was wearing only a thin jacket, no hat nor scarf, and whose face was the pink of those who have only just arrived and still describe the cold as *exhilarating*. The man nodded slowly, for no one made quick movements in this town, turned around and fetched a key from the board behind.

'Twenty-nine,' he said. 'Up the stairs, first left, first right. Towels in the bathroom, breakfast's extra.'

'Right,' said Sergeant, taking the key and his bag. He headed for the stairs, whistling as he went. The man behind the counter watched him, wincing slightly at the whistle. Sergeant felt the man's eyes on his back. That's right, the young man thought, That's right.

In truth, Sergeant was not as young as he appeared. He had a method for preserving his youthful appearance which he was not planning on divulging, even to his fiancée, Elaine, whom he would marry the following June. This secret had been handed down to him by an elderly aunt who favoured him over her twelve other nephews and nieces.

Calling him aside during a family occasion when he had only just entered into the age at which boys' faces harden and bristle, she passed the knowledge to him as it had been passed to her by her grandmother. After she had told it, she moved closer to the boy. 'You will not live longer,' she warned, 'but you will appear to be twenty years less than you are.' Looking herself not a day past forty, though she had in fact overtaken sixty at last birthday, she took hold of his shoulders. 'This can be both blessing and curse, you understand me?' Sergeant, who didn't, just nodded. He saw only the blessing, and his aunt saw that he would fall because of it and she could do nothing to prevent the fall. The boy kissed his aunt on the cheek, and if he noticed that her skin was icy despite the late August heat, the thought passed straight through and out of his mind. With her aged eyes, that told anyone who looked closely at her smooth, soft face that mischief was at work here, his aunt watched him run back to his cousins and sighed for what was to come.

In the small hotel room, Sergeant shut the curtains and unpacked his bag, hanging up two dark suits in the wardrobe and placing the large metal box under the single bed. Then he sat down in the chair and smiled a smile that a room so small, dark and musty rarely saw. He would soon find out if all the years of preparation had come to something, or to nothing. He would soon know if he would make his fortune.

Struck with a sudden hunger, he stood, put on his jacket, and went to discover what the town could offer. He ran down the stairs, whistling as he jogged through the lobby.

The man at the hotel desk watched him, annoyed.

'Who is he?' he asked the chambermaid who had come to request the next day off work due to her father's worsening overnight.

'Don't know,' said the chambermaid, whose father's condition had not changed but whose boyfriend demanded she come out with him to the woods. Because she was more afraid of her boyfriend than of losing her job she was doing what she had to do.

'No word of how long he's staying, no idea at all,' mumbled the man.

'That's alright then?' asked the chambermaid.

'Go on with you,' the old man said, and slowly bent himself to reach the employee book on the lower shelf and log the change.

The next morning, Sergeant rose late. The sun was already high in the sky, although the town did not feel it, feeling only the chill and the damp. Sergeant smelled the air and nodded. He washed and dressed quickly and then withdrew the case from underneath the bed. From the case he took a large ball, the size of a small watermelon, but not so heavy that he could not hold it comfortably in both hands. The ball was a made of metal, silver in one light, golden in another, and despite the dullness of the day, it shone. It was an object that held your gaze, letting you look at nothing else.

With the ball in both his hands, Seargant stood by the window. Setting the ball on the sill, he opened the window wide and the cold air flowed in. He picked up the globe which had begun to tremble.

'That's right,' he said, 'that's right,' and, holding it on the boundary between inside and out, he closed his eyes and breathed a deep breath.

A substance began to stream from the sides of the ball, yellow in colour and with a texture like steam. It flowed out and down, towards the town and over the town, until the whole area was covered in a fine buttercup mist. Sergeant, eyes still closed, turned his face upwards to feel the heat on his cheeks.

In the woods, the chambermaid's boyfriend was fumbling with the fifth layer of her woollens while she lay there, her head on frozen leaves, next to her the hat that had fallen off as he manhandled her. She felt nothing. She stared at the frost-covered tree stump beside her. Her boyfriend grunted as he made progress, and inside herself the chambermaid sighed. She could not even cry, because she knew that her tears would freeze as they ran down.

Something stroked her cheek and she started. *He* was still down the other end engaged in exploratory digging. There was no one there. But her cheek was warm. Slowly she realized that it was not just her face—her body started to experience a sensation she had never had outside of her heavily-blanketed bed.

She was hot.

She was feeling hot.

She sat up.

'Stop!' she said. 'Can't you feel it!' He grunted but broke off from what he was doing to stare at her. The chambermaid struggled to her feet and began pulling off her clothes. 'I'm hot! I'm hot!' she cried. She threw the clothes at her boyfriend, who fell backwards under the weight of the wool, and the chambermaid, wearing only her underwear, started running towards the town.

Sergeant walked out into the street. The townspeople were converging, wanting to know what it was, this strange new sensation. People were disrobing, large jumpers and cardigans, socks and thick tweed trousers lying on the ground.

'Bloody boiling . . . !'
'Something's not right, I'm . . .'
'Have to take these off, it's . . .'
'Never in all my days, have . . .'
'Praise be . . .

. . . to me, thought Sergeant. Praise be to me, and he stepped into the middle of Main Street and started walking through the crowds.

No one knew what was happening, why the cold was leaving their frozen town. One man had an idea. Wearing only his vest and underpants, he left the hotel desk unattended and made his way slowly upstairs, unaccustomed to the sweat on his forehead and spreading across his chest. At Room 29, turned his master key in the lock and went in. It took him a few minutes to find the ball, which was back in its case under the desk. He held it in both hands, and he felt coursing through him the heat from a thousand warm souls, the breath of a thousand people living their days in the sun in every country Sergeant had travelled to for the gathering of his substance. The old man's lips, so long frozen into an expression of disdain, began to turn upwards. Inside him ice was shifting. He walked towards the open window and held the ball outwards.

'See!' he cried, but his voice was weak and no one heard him, just as they did not see the yellow streaming again over their town.

Sergeant saw.

Sergeant saw that there was more where it should have stopped, more when there was already plenty.

'No!' he cried, but he was too far away. He began running towards the hotel but the heat was inside him and his legs moved slower and slower. He stopped and sank to his knees. He put his hands to his face and felt the wrinkles crawling across his brow. Moving upwards, he found his hair receding away from him. His fingers began to bend and stiffen. Age crept into every muscle, every joint, not just the twenty years that he had pushed deep inside of him, but twenty and twenty and twenty more, for the damage he had done to his nature. His face crumpled into folds of old skin, his vision began to blur, and still the golden haze flowed across the town.

Sergeant cried out for help, but no one heard him. The townspeople, naked now, were dancing in the glow of the warm yellow light, and they danced around him in their joy, whirling and singing, as he sank slowly down towards the ground.

ACKNOWLEDGEMENTS

Writing is a process begun alone but involving many people along the way. Thank you Dina, Devorah, Lisa, Ally, Louise, Mazzy, Vanessa, Sara, Elaine, Elizabeth, Susannah, Mel, Zoe, Kuzhali, the Fiction Workhouse, the WriteWords FF1 group, Ilene, Aloma, Batnadiv, Nadia, Leeora, for taking the time to read my writing, share your work with me, and offer invaluable critique. Thank you to Jeremy Osborne at Sweet Talk for his constant support of my writing and of the short story in general. Thank you to all my fellow writer/bloggers who share their thoughts about writing and reading, creating an online community of like minds. Thank you to all my *Short Review* reviewers for making sure short story collections get their turn in the spotlight, teaching me so much through your thoughtful and insightful reviews.

Thank you to *New Scientist* for the fascinating articles, which provided the inspiration for stories in this collection.

An enormous thank you to the publications that chose to publish some of the stories, or versions of the stories, in this book: we short story writers would be lost without you: *Route* ('The White Road', 'On a Roll'; both were also broadcast on BBC Radio 4's Afternoon Reading); the *Frogmore Press* ('Heavy Bones'); the *Binnacle* ('The Hand'); the *Steel City Review* ('Space Fright'); *Creating Reality* ('Plaits', 'Mugs'); *LauraHird.com* ('Sunspots'); *Every Day Fiction* ('Go Away', 'Fish-filled Sea'); *Belvedere Writers' Club* ('Evie and the Arfids'); *RethinkDaily podcast* ('The Angel in the Car Park'); *Magazine Minima* ('Heart'); *Transmission* magazine ('Express'); *Front&Centre* ('Exchange Rates'); the *Entelechy Review* ('My Name is Henry'), and *Riptide* ('North Cold').

Thank you to the wonderful teachers I have had over the years, who have all shown me what a short story can look like and generously given me tools to create my own.

Thank you to Zac and Cleo for purring gently beside me while I write. Thank you to my parents, Anita and Brian, who, despite wanting me to pursue a conventional career, supported my deviation from the norm and, along with my stepmother, Carole, are always interested in reading my stories and delighted in my success. Thank you to my brother, Nick, for the conversations about the nature of creativity. Most of all, thank you to James, always there for me, understanding my need to shut the door and disappear into the world of my characters, supporting and feeding me in so many ways.